EARTH SCIENCES IN THE 21ST CENTURY

NEW DEVELOPMENTS IN MAGNETOHYDRODYNAMICS, PLASMA AND SPACE RESEARCH

EARTH SCIENCES IN THE 21ST CENTURY

Additional books in this series can be found on Nova's website under the Series tab.

Additional E-books in this series can be found on Nova's website under the E-book tab.

SPACE SCIENCE, EXPLORATION AND POLICIES

Additional books in this series can be found on Nova's website under the Series tab.

Additional E-books in this series can be found on Nova's website under the E-book tab.

EARTH SCIENCES IN THE 21ST CENTURY

NEW DEVELOPMENTS IN MAGNETOHYDRODYNAMICS, PLASMA AND SPACE RESEARCH

STEPHEN TALTON
AND
PATRICK F. THEBA
EDITORS

Nova Science Publishers, Inc.
New York

Library of Congress Cataloging-in-Publication Data

New developments in magnetohydrodynamics, plasma, and space research / editors, Stephen Talton and Patrick F. Theba.
p. cm.
Includes bibliographical references and index.
ISBN 978-1-61942-078-6 (hardcover)
1. Magnetohydrodynamics. 2. Plasma dynamics. I. Talton, Stephen. II. Theba, Patrick F.
QC718.5.M36N49 2011
538'.6--dc23

2011042416

Published by Nova Science Publishers, Inc. † New York

CONTENTS

PREFACE

This book presents and discusses current research in the study of magnetohydrodynamics. Topics discussed include consideration of the problems of gravitational optics and dark matter based on the crystal model of the vacuum; origin of the Saturn rings; high speed flowfield analysis for satellite launch vehicle and reentry capsule; MHD flow of a micropolar fluid with asymmetric wall tempreatures; Na and Hansen's procedure for solving the MHD flow around a cylinder in cross electric and magnetic fields and a unified approach to analytical solution of an unsteady MHD generalized couette flow with transpiration.

Chapter 1 – In presented chapter the authors try to consider problems of the gravitational optics and dark matter developing from the crystal model for the vacuum. How it is follows from consideration it enables to describe both electromagnetic waves and spectrum of elementary particles from the unified point of view. Two order parameters – a polar vector and an axial vector - had to be introduced as electrical and magnetic polarization, correspondingly, in order to describe dynamic properties of vacuum. Vacuum susceptibility has been determined to be equal to the fine structure constant α. Unified interaction constant g for all particles equal to the double charge of Dirac monopole has been found ($g = e/\alpha$, where e charge electron). The fundamental vacuum constants are: g, α, parameters of length ξ_e, ξ_n and parameters of time τ_e, τ_n for electron and nucleon oscillations, correspondingly. Energy of elementary particles has been expressed in terms of the fundamental vacuum parameters, light velocity being equal to $c = \xi_e/\tau_e = \xi_n/\tau_n$. The term mass of particle has been shown to have no independent meaning. Particle energy does have physical sense as wave packet energy related to vacuum excitation. Exact equation for particle movement in the gravitational field has been derived, the equation being applied to any relatively compact object: planet, satellite, electron, proton, photon and neutrino. The situation has been examined according to the cosmological principle when galaxies are distributed around an infinite space. In this case the recession of galaxies is impossible, so the red shift of far galaxies' radiation has to be interpreted as the blue time shift of atomic spectra; it follows that zero-energy, and consequently electron mass are being increased at the time. Since physical vacuum has been existed eternally, vacuum parameters can be either constant, or oscillating with time. It is the time oscillation of the parameters that leads to the growth of electron mass within the last 15 billion years and that is displayed in the red shift; the proton mass being decreased that is displayed in planet radiation.

Chapter 2 – For the first time the role of superconductivity of the space objects within the Solar system located behind a belt of asteroids is considered. Observation of experimental data for the Saturn's rings shows that the rings particles may have superconductivity. Theoretical electromagnetic modeling demonstrates that superconductivity can be the physical reason of the origin of the rings of Saturn from the frozen particles of the protoplanetary cloud. The rings appear during some time after magnetic field of planet appears. It happened as a result of interaction of the superconducting iced particles of the protoplanetary cloud with the nonuniform magnetic field of Saturn. Finally, all the Kepler's orbits of the superconducting particles are localizing as a sombrero disk of rings in the magnetic equator plane, where the energy of particles in the magnetic field of Saturn has a minimum value. Electromagnetism and superconductivity allow us to understand why planetary rings in the solar system appear only for the planet with the magnetic field after the belt of asteroids where the temperature is low enough and why there are no rings for the Earth, and many other phenomena.

Chapter 3 – The chapter presents numerical and experimental analyses of satellite launch vehicle and reentry capsule at high-speeds. Numerical simulations of free jets emanating from nozzle, impinging of supersonic jets on jet deflector, flowfield over bulbous payload shroud in the transonic and supersonic Mach number range, drag reducing spike attached to blunt body, prediction of angle of attack from the flight data of launch vehicle and depressurization of heat shield during ascent phase of flight. Influences geometrical parameters of various types of reentry capsules on aerodynamic drag are studied at various trajectory points. Results of computational fluid dynamics are compared with the experimental and flight data and exhibit fairly good agreement between them. The reliable estimation of aerodynamic problem associated with launch vehicle can be made in conjunction with computed flow visualizations.

Chapter 4 – This chapter presents the effect of MHD of micropolar fluids with asymmetric wall temperatures in a vertical wall. The transformed boundary layer governing equations are solved by the Matlab software. The dynamic of micropolar fluids has been a popular area of research. As the fluid consist of randomly oriented molecules, and as each volume element of the fluid have translated as well as rotation. The analysis of physical problems in these fluids has revealed several interesting phenomena, which are not found in Newtonian fluids. Increasing the Magnetohydrodynamics parameter tends to decrease the velocity profile while the micro rotation increase as the Magnetohydrodynamics increases in vertical walls.

Chapter 5 – An analysis of the magneto hydrodynamic (MHD) flow of a liquid metal around a cylinder of arbitrary electrical conductivity, under the influence of a magnetic field perpendicular to an electric current has been presented. Na and Hansen's procedure as a procedure for reducing the number of independent variables is applied. Finally, the obtained ordinary differential equations with the appropriate corresponding conditions are solved analytically. The flow field, the induced magnetic field, and the resulting forces on the cylinder are calculated.

Chapter 6 – The mathematical model describing the unsteady generalized Couette flow with porous boundaries of viscous incompressible, electrically conducting fluid is solved analytically for general nature of time-dependent pressure gradient and initial condition. For the general nature of motion of upper boundary, unified expression has been derived for the fluid velocity corresponding to the cases of magnetic field fixed relative to the fluid or to the upper boundary.

Chapter 7 – A study is made to analyze the unsteady MHD (magneto-hydrodynamic) two phase flow in a horizontal channel filled with conducting fluids of different physical properties in the presence of transverse magnetic field for short circuit case. It is assumed that the conducting fluids are having constant transport properties. The two regions are coupled by equating the velocity and shear-stress at the interface. Using the Green's function method, expressions for velocities for both phases are obtained for general class of time dependent movement of boundaries or sudden change in pressure gradient or both. As a special case, expressions for velocities in both phases are obtained due to sudden change in pressure gradient.

Chapter 8 – Resonance exhibits a controlled thermonuclear fusion reaction of the Sun in vacuum permeated by a magnetic mirror with a decisive importance to a driving force. An abundance of helium with the Sun is exposed to a nuclear bombardment with ionized hydrogen with reference to a chain reaction so that hydrogen bomb blasting on the surface of the Sun is reliably exhibited. With reference to a driving force it is rigorously stated that a laser radiation communicates ultraviolet radiation and X – ray emission at the resonant level as the magnetic field grows towards the resonant level with reference to a radio frequency accelerator with the Sun.

Chapter 9 – This chapter is concerned with the steady mixed convection boundary layer flow of viscous incompressible electrically conducting fluid past a semi infinite magnetized vertical porous plate with uniform transpiration and variable transverse magnetic field along the surface. The equations governing the flow, magnetic and temperature field are reduced to dimensionless convenient form using the free variable transformations and solved numerically by using finite difference method. Effects of the physical parameters, such as, Pr, S, and λ on the numerical values of skin friction coefficient, $Re_x^{1/2}Cf_x$, heat transfer coefficient, $Re_x^{-1/2}Nu_x$, and coefficient of magnetic intensity, $Re_x^{1/2}Mg_x$ against ξ, the local transpiration parameter are presented in tabular form and the velocity profile, temperature profile and the profile of transverse component of magnetic field are shown graphically. Asymptotic solutions are also obtained for smaller and larger values of ξ and compared with the entire values of ξ.

Versions of these chapters were also published in *Journal of Magnetohydrodynamics, Plasma and Space Research*, Volume 15, Numbers 1 and 2, published by Nova Science Publishers, Inc. They were submitted for appropriate modifications in an effort to encourage wider dissemination of research.

In: New Developments in Magnetohydrodynamics …
Editors: Stephen Talton and Patrick F. Theba

ISBN: 978-1-61942-078-6
2012 Nova Science Publishers, Inc.

Chapter 1

CONSIDERATION OF THE PROBLEMS OF GRAVITATIONAL OPTICS AND DARK MATTER BASED ON THE CRYSTAL MODEL OF THE VACUUM

Evgeny V. Chensky*

Modern Science Institute, SAIBR. Moscow, Russia

ABSTRACT

In presented chapter we try to consider problems of the gravitational optics and dark matter developing from the crystal model for the vacuum. How it is follows from consideration it enables to describe both electromagnetic waves and spectrum of elementary particles from the unified point of view. Two order parameters – a polar vector and an axial vector - had to be introduced as electrical and magnetic polarization, correspondingly, in order to describe dynamic properties of vacuum. Vacuum susceptibility has been determined to be equal to the fine structure constant α. Unified interaction constant g for all particles equal to the double charge of Dirac monopole has been found ($g = e/\alpha$, where e charge electron). The fundamental vacuum constants are: g, α, parameters of length ξ_e, ξ_n and parameters of time τ_e, τ_n for electron and nucleon oscillations, correspondingly. Energy of elementary particles has been expressed in terms of the fundamental vacuum parameters, light velocity being equal to $c = \xi_e/\tau_e = \xi_n/\tau_n$. The term mass of particle has been shown to have no independent meaning. Particle energy does have physical sense as wave packet energy related to vacuum excitation. Exact equation for particle movement in the gravitational field has been derived, the equation being applied to any relatively compact object: planet, satellite, electron, proton, photon and neutrino. The situation has been examined according to the cosmological principle when galaxies are distributed around an infinite space. In this case the recession of galaxies is impossible, so the red shift of far galaxies' radiation has to be interpreted as the blue time shift of atomic spectra; it follows that zero-energy, and consequently electron mass are being increased at the time. Since physical vacuum has been existed eternally, vacuum parameters can be either constant, or oscillating with time. It is the time oscillation of the parameters that leads to the growth of electron mass within the last

* Email address: +7(926)592-6066, echensky@yandex.ru

15 billion years and that is displayed in the red shift; the proton mass being decreased that is displayed in planet radiation.

INTRODUCTION

The science about cosmology has been in rather difficult situation in recent years. On one hand, observations of star dynamics in galaxies and of galaxies in clasters show substantial deviation of rotation velocities from Kepler's law; this proves the existence of additional matter (dark matter) which participates in gravitational interaction [1-3]. On the other hand, more careful examination of the red shift in the nearer space at the distances of 105-107 light years as well as observation of supernova outburst [4-5] show that velocity of the Universe expansion increases with time, and this in turn requires introduction of additional dark energy with anti-gravitational properties. Thus, a contradiction arises. Practically, in one and the same point it is necessary to introduce both dark matter creating additional gravitational field and dark energy having anti-gravitation. Since there is no doubt about the facts above, their interpretation must be revised.

At the present time there are two mutually exclusive points of view. First, despite very distinctive spatial non-homogeneity of matter, observations show that at the distances of about 10^9 light years (cell of homogeneity) matter is distributed in the space quite homogeneously. Besides, the cosmological principle suggests that these homogeneous cells should cover the entire infinite space. Second, the red shift discovered by Hubble, which he interpreted as Doppler's principle related to the galaxies expansion, made Friedman's model of expanding Universe quite necessary. From Hubble's empirical law that determines dependence of velocity of galaxies on the distance $v = Hr$, we can suppose existence of a singularity at a certain time. Since velocity of expansion of the galaxies cannot exceed the light velocity c, it follows from the relation $c = HR = HcT_0$, that there is quite a definite size of the Universe growing with time $R = cT_0$, here – T_0 is the singularity offset counted from the present moment; Hubble's constant equal to $H = 1/T_0$ decreases with time; however, observations show, that the value H, on the contrary, increases with time.

If we interpret the existence of a singularity as a Big Bang, we have to bear in mind that the explosion is a phase transition from a metastable state into another more stable state accompanied with release of energy. Before the phase transition, this energy is homogeneously distributed around the space. They sometimes say: explosion power is equivalent to e.g. one kilogram of trotyl; it is obvious that two kilograms of trotyl give off right twice as much energy as one kilogram does. Besides, the phase transition does not begin with the singularity but with the nucleation of a new phase whose size exceeds the critical radius. In this case energy is released in accordance with broadening the new phase at the expense of the phase edge motion. Since the average energy density of the entire matter in vacuum is approximately 0,008 erg/m^3, this very energy should be released at the phase transition of each cubic meter of vacuum. It is difficult to imagine, however, that electrons and protons could be created out of this homogeneously distributed in space energy, and, besides, in exactly equal quantities. An explosion of a hydrogen bomb in vacuum can serve as a model of a hot Universe. The hydrogen bomb is a local object in a metastable state. There is a mixture of light and heavy nuclei under the temperature of several million degrees at the moment of detonation. According to D'Alambert equation, the electromagnetic pulse and the neutrino pulse will start to disperse with the light velocity. Following electromagnetic pulse relativistic electrons will fly and then light, and heavy nuclei. In a second, the electromagnetic pulse will reach the Moon area and nothing will stay at the point of explosion. Thus, the examined case is also far from the Friedman's model of expanding Universe.

In order to somehow reconcile the model of the infinite matter distribution in space with that of the expanding Universe, Milne offered the following reasoning [6]. If we mentally specify a sphere of a definite size in a matter homogeneously distributed around an infinite space, then external layers of the sphere due to their spherical symmetry have no influence on the sphere dynamics. Therefore, we can ignore the external layers and consider the Universe as a sphere of a definite size that precisely coincides with the Friedman's model. However, this statement is a mistake. The thing is that with matter being homogeneously distributed about the entire infinite space, the gravitational potential follows the condition of the translational invariance: $U(\vec{r})=const$. We may consider this constant to be equal to zero, therefore, a gravitational potential only arises at deviation of a matter distribution from an average value. For that reason the equation for the potential can be written as follows:

$$\Delta U(\vec{r})=4\pi\gamma\left(\rho(\vec{r})-\rho_0\right) \tag{1}$$

Here ρ_0 is an average density of matter. From equation (1) we can see that it is not necessary to search for dark energy as the density is both the gravitating and the anti-gravitating matter in the form of $\rho(\vec{r})$ and ρ_0.

On the other hand, if we mentally specify a sphere of radius R with the density of matter ρ_0 and ignore the external matter, we come to another equation for the potential:

$$\Delta U(\vec{r})=4\pi\gamma\begin{cases}\rho_0, & r<R;\\ 0, & r>R.\end{cases}$$

This equation has the following solutions:

$$U(r)=-2\pi\gamma\rho_0R^2\left(1-\frac{r^2}{3R^2}\right),\quad r<R;$$
$$U(r)=-\frac{4\pi\gamma\rho_0R^3}{3r},\quad r>R. \tag{2}$$

Similar expressions can be used for determining gravitational potentials of planets, stars and galaxies in a form of the sum of the potentials of stars with their specific location. However, for the scales comparable with the size of homogeneity cell and bigger, we come to an obviously non-physical result: the potential in any arbitrary point depends on the radius of a sphere which we mentally specify out of the entire infinite space. Thus, any result depending on the mentally specified radius of the sphere, including the radius of the visible part of the Universe, is physically incorrect.

For instance, we can determine the circular orbital velocity v_1 for the Universe of radius R on the sphere surface from the equality of centripetal and centrifugal forces:

$$R\dot{\varphi}^2 = \frac{\partial U(r)}{\partial r}\Big|_{r=R} = \frac{4\pi\gamma\rho_0 R}{3}; \quad v_1 = R\dot{\varphi} = R\sqrt{\frac{4\pi\gamma\rho_0}{3}}$$

If v_1 is equal to the light velocity c, we obtain the following expression for the critical matter density in the Universe:

$$\rho_c = \frac{3c^2}{4\pi\gamma R^2} = \frac{3H^2}{4\pi\gamma}.$$

This corresponds to the condition $R = r_g$ when r_g is a gravitational radius. Therefore, with the definite choice for R we may come to the conclusion that the Universe is a black hole, while, as it follows from the cosmological principle at the scales comparable with the radius of the visible part of the Universe, the gravitational potential has no specific features and its average value is zero.

The same situation takes place when we consider the influence of a pressure on the dynamics of the expanding Universe. For instance, if we take a big vessel with a gas, mentally specify a sphere of radius R in it, and ignore the gas surrounding the sphere, we can state that the gas will broaden and get cool at the expense of the internal pressure. This may remind the model of the expanding Universe. Remember, however, that the specified sphere is surrounded with the same gas at the same pressure; that is why there will be neither broadening nor cooling. Thus, for the infinite Universe both an average gravitational potential and an average pressure are constant; besides, since the expanding dynamics is influenced by the equal to zero gradients of these variables, there cannot be neither expanding nor compression. An infinite system can only stratify according to the energy density and we really observe this stratification on giant scales from the value less than 10^{-9}erg/cm^3 for an inter-galaxy space to the value over 10^{39} erg/cm^3 for nuclear energy.

Nevertheless, within the frames of the cosmological principle there is a problem, the so called photometric paradox. The thing is that at present time when stars and galaxies radiate light in the entire infinite space, we can introduce an average luminosity L of a unit volume, provided that the densities of a luminous flux intensity at the distance r from a single volume is equal to $j = L/4\pi r^2$. The integral over the sphere of radius R gives the total flux intensity equal to $J = RL$; it follows that with R approaching infinity the flux intensity must approach infinity as well. Practically, however, we see rather a low sky luminosity. This is the photometric paradox.

In fact, by calculating the intensity, we must take into consideration the retardation effects. The flux that comes to a certain point $(r = 0)$ at a certain time $(t = 0)$ radiates at different moments depending on the distance:

$$J = \int j\left(t = -\frac{r}{c}\right)d\bar{r} = c\int_{-\infty}^{0} L(t)dt \qquad (3)$$

Expression (3) shows that the flux coming from the deep Universe will be finite if $L(t)$ at longer t decreases faster than $1/t$.

Besides, we can divide the entire flux observed at any point of the infinite space into two parts: the flux J_{vis} of a visible part of space $R = cT_0$, $T_0 \sim 15 \cdot 10^9$ years and the relict flux J_{rel} radiating from the spots with $r > R$:

$$J = J_{vis} + J_{rel} = \sum_{n=1}^{N} J_n + \int_{-\infty}^{-T_0} L(t')dt'$$

Here summing was carried out over a countable number of galaxies in the visible part of the Universe. Thus, from the expressions given above it follows that the Universe must be non-stationary, not due to an expansion of galaxies', but at the expense of a variation of physical vacuum parameters. Since the relict radiation corresponds to the temperature 3^0K, the Universe had such a temperature long ago. The one but not the only feature of a non-stationarity is the red shift of atomic spectra that we can interpret as the blue temporal shift of both characteristic Bohr energy and all atomic energy levels correspondingly, at the expense of variation of physical vacuum parameters. Observations show that the characteristic Bohr frequency depends on time and increases with time. By introducing a frequency of an arbitrary atomic level, we obtain the following expression for the Hubble's constant:

$$\frac{d}{dt}\left(\frac{\omega(t)}{\omega(t=0)}\right) = H(t) \quad (4)$$

Both $\omega(t)$ and $H(t)$ are monotonously increasing functions. The latest observations of the flashes of far supernova [4-5] show temporal growth of *H*. It is senseless to explain this situation using space-time properties.

Speaking about space-time properties is quite the same as judging about wine quality by the curvature of a bottle surface. Dilettantes are often attracted by the appearance of the vessel, while connoisseurs pay attention to its contents, conservation conditions, and temporal changes. We should regard space like a vessel with the only feature: its volume is infinite. Its internal properties are to be discussed.

Hidden Parameters of Vacuum

We should proceed from the experimental fact that the energy and the pulse of any elementary particle are:

$$\varepsilon_k = \hbar\omega_k; \qquad \vec{p} = \hbar\vec{k} \quad (5)$$

here ω_k – frequency for electron, proton, photon and neutrino, correspondingly, we expressed as follows:

$$\omega_{ek} = \sqrt{\omega_{e0}^2 + c^2k^2}; \qquad \omega_{pk} = \sqrt{\omega_{p0}^2 + c^2k^2}; \qquad \omega_{\gamma,\nu;k} = ck \quad (6)$$

The unified formula for the energy of any elementary particle points to the existence of the universal interaction for fields related to each particle. Besides, the two oscillation branches with the energy gap observed in the excitation spectrum prove an existence of a certain set of discrete oscillators whose interaction causes normal oscillations with frequencies ω_k. In fact, we can represent vacuum as a crystal object of a cubic or hexagonal symmetry with a very small lattice period, much less than 10^{-26} cm. We can estimate the upper limit of a lattice period by the maximum particle energy in cosmic rays equal to 10^{21} eV that corresponds to the wave vector of 10^{26} cm^{-1}. The vacuum ground state is the equilibrium position of all oscillators; these are the points of equilibrium forming crystal lattice related to the absolute coordinate system. Under the deviation of an oscillator from the equilibrium position, a dipole moment arises. For the scales exceeding the lattice period we can introduce a macroscopic order parameter as an electric polarization of vacuum:

$$\vec{P}(\vec{r}) = \frac{4\pi}{\delta V}\sum_i \vec{d}_i .$$

Suppose, there are two branches of normal oscillations of field $\vec{P}$ that we can call electron and nucleon modes. The Hamiltonian for electron and nucleon modes written in the unified form, is:

$$H = \frac{1}{8\pi}\int\left(\tau_{e,n}^2 \dot{\vec{P}}^2 + \xi_{e,n}^2\left(\nabla\vec{P}\right)^2 + \sigma\vec{P}^2\right)d\vec{r} \qquad (7)$$

For electronic and nucleonic parts, we introduced the parameters of time τ_e, τ_n and length ξ_e, ξ_n that characterize the kinetic and gradient energy of the fields. Besides, we introduced a dimensionless parameter of an elastic coefficient σ corresponding to the reciprocal susceptibility common to both modes. These are the latent parameters of vacuum and the available experimental data are sufficient to determine them.

By using the minimal action principle for the Lagrange function equal to the difference of the kinetic energy - the first member of expression (7), and the potential energy – the second and the third members of (7), we obtain the equation of motion for six independent normal oscillations $P_{ex}, P_{ey}, P_{ez}, P_{nx}, P_{ny}, P_{nz}$:

$$\left(\sigma + \xi_{e,n}^2\left(\frac{\partial^2}{\partial x^2} + \frac{\partial^2}{\partial y^2} + \frac{\partial^2}{\partial z^2}\right) - \tau_{e,n}^2\frac{\partial^2}{\partial t^2}\right)P_{e,n,x,y,z} = 0 \qquad (8)$$

By setting up the following solutions:

$$P_{x,y,z} = a_{x,y,z}\exp\left(i\left(\vec{k}\vec{r} + \omega t\right)\right), \qquad (9)$$

we obtain the spectrum for normal oscillations:

$$\omega_{e,n;k} = \pm\sqrt{\omega_{e,n;0}^2 + c^2k^2};$$

$$\omega_{e,0} = \frac{\sqrt{\sigma}}{\tau_e}; \qquad \omega_{n,0} = \frac{\sqrt{\sigma}}{\tau_n}; \qquad c = \frac{\xi_e}{\tau_e} = \frac{\xi_n}{\tau_n}. \tag{10}$$

Therefore, we can represent physical vacuum as some coherent state with the natural frequency standards in the form of homogeneous polarization oscillations about an absolute coordinate system

$$P_{e,n;x,y,z}(\vec{r},t) = a_{e,n;x,y,z}\exp(i\omega_{e,n;0}t)$$

with the absolute time, homogeneous around the entire space $t(\vec{r}) \equiv t_{abs}$.

The situation, however, becomes more complicated, since the electrical vacuum polarization generates the following electric charge:

$$div\vec{P} = -4\pi\rho_e. \tag{11}$$

Here ρ_e is the electric charge density, while the polarization is determined by both electron and nucleon modes $\vec{P} = \vec{P}_e + \vec{P}_n$. This results in an additional long-range Coulomb interaction between the normal oscillations $P_{ex}, P_{ey}, P_{ez}, P_{nx}, P_{ny}, P_{nz}$

$$U = \frac{1}{2}\iint \frac{\rho_e(\vec{r})\rho_e(\vec{r}')}{|\vec{r}-\vec{r}'|}d\vec{r}d\vec{r}' \tag{12}$$

For simplicity, we consider normal oscillations inside the electronic modes. We dimensionlize coordinates and time. We express new variables like this: $t \Rightarrow t/\tau_e$; $r \Rightarrow r/\xi_e$; velocity being in terms of $c = \xi_e/\tau_e$. It makes sense to specify a dimensional value for the electric polarization in the terms of the electron charge:

$$\vec{P} \Rightarrow \frac{e}{\xi_e^2}\vec{P},$$

after that the electron field action reduces to form:

$$S = \frac{e^2}{8\pi c}\int dt\int d\vec{r}\left\{\dot{P}_i\dot{P}_i - \frac{\partial P_i}{\partial x_j}\frac{\partial P_i}{\partial x_j} - \sigma P_iP_i - \frac{1}{4\pi}\int\frac{\dfrac{\partial P_i(\vec{r},t)}{\partial x_i}\dfrac{\partial P_j(\vec{r}',t)}{\partial x_j'}}{|\vec{r}-\vec{r}'|}d\vec{r}'\right\} \tag{13}$$

Here i and j run over x, y, z and we carried out summation over repeated indices. By varying action S over the values $P_i, \dot{P}_i, \partial P_i/\partial x_j$, we come to the following system of the integral-differential equations:

$$\left(\frac{\partial^2}{\partial t^2}-\Delta+\sigma\right)P_i(\vec{r},t)+\frac{1}{4\pi}\int\frac{(x_i-x_i')}{|\vec{r}-\vec{r}'|^3}\frac{\partial P_j(\vec{r}')}{\partial x_j'}d\vec{r}'=0 \quad (14)$$

Consider the solutions in the form of plane waves:

$$P_i(\vec{r},t)=P_i\exp\left(i\left(\vec{k}\vec{r}+\omega t\right)\right) \quad (15)$$

For plane waves, equations (14) reduce to the form:

$$\left(\sigma+k^2-\omega^2\right)P_i+\frac{k_ik_j}{k^2}P_j=0 \quad (16)$$

By making the determinant of the equations (16) equal to zero, we obtain the oscillation spectrum:

$$\begin{vmatrix}\left(\sigma+k^2+\frac{k_x^2}{k^2}-\omega^2\right) & \frac{k_xk_y}{k^2} & \frac{k_xk_z}{k^2}\\ \frac{k_xk_y}{k^2} & \left(\sigma+k^2+\frac{k_y^2}{k^2}-\omega^2\right) & \frac{k_yk_z}{k^2}\\ \frac{k_xk_z}{k^2} & \frac{k_yk_z}{k^2} & \left(\sigma+k^2+\frac{k_z^2}{k^2}-\omega^2\right)\end{vmatrix}=0 \quad (17)$$

Equation (17) transforms to

$$\left(\sigma+k^2-\omega^2\right)^2\cdot\left(\sigma+1+k^2-\omega^2\right)=0 \quad (18)$$

Thus, from (18) we obtain the normal spectrum of the oscillations; from equations (16), we obtain the form of the oscillations:

$$\omega_k=\pm\sqrt{\sigma+1+k^2};\ \ k_yP_x=k_xP_y,\ \ k_zP_x=k_xP_z,\ \ k_zP_y=k_yP_z;$$
$$\omega_k=\pm\sqrt{\sigma+k^2};\qquad k_xP_x+k_yP_y+k_zP_z=0. \quad (19)$$

Expressions (19) allow the definition of the general properties of the normal oscillations for vacuum fields linked by the long-range Coulomb interaction. The Laplacian operator in equations (14) requires that the polarization components be eigenfunctions of this operator:

$$\Delta P_i=\beta^2P_i \quad (20)$$

The result of the Coulomb interaction is that the oscillations of the polarization are divided into two classes: longitudinal $\vec{P}_1$ with $rot\vec{P}_1=0$ and lateral $\vec{P}_2$ with $div\vec{P}_2=0$, according to the Helmholz theorem $\vec{P}_1 = \nabla\Phi$; $\vec{P}_2 = rot\vec{A}$, here Φ and $\vec{A}$ are scalar and vector potentials. Longitudinal oscillations provide a depolarizing electric field $\vec{E} = -\vec{P}_1$, which meets the following condition:

$$div\vec{E} = 4\pi\rho_e.$$

For lateral (transverse!) oscillations, the depolarizing field equals to zero. As a result, the frequencies for longitudinal and lateral oscillations are different.

The problem, however, is that for linear homogeneous differential equations we may take into consideration both eigenfunctions and eigenvalues, while the amplitude of the eigenfunctions remains arbitrary. Suppose, an eigenfunction specifies the configuration of the excitation; though the excitation energy and pulse are the integrals of motion, and yet they can have arbitrary meanings. Nevertheless, in practice we can see that energy of any excitation has quite a definite meaning both for light quantum and for any elementary particle. Therefore, within the framework of homogeneous equations it is impossible to realize the origin and the physical meaning of the Planck constant.

For linear systems, the amplitude of oscillations turns out to be quite definite under the external force; then we can express the solution by means of the Green function, which meets the homogeneous equation and has quite definite amplitude. Non-homogeneous equations are necessary for the following reasons. We know from the theory of many-body systems that, if a system consists of discrete particles, the correlation effects substantially decrease the ground state, and local states such as polarons can occur. Therefore, we pass to consideration of the ground state taking into account correlation effects.

From an endless number of particles forming a crystalline vacuum state we examine one particle as a point unit source $Q_e = \delta(\vec{r})$, which generates longitudinal electric field defined by equations:

$$div\vec{E}_0 = 4\pi\delta(\vec{r}); \; rot\vec{E}_0 = 0; \; \vec{E}_0 = \frac{\vec{r}}{r^3}.$$

Thereafter, we can write the interaction energy of the point source with vacuum fields as follows:

$$U_0 = -\frac{g}{4\pi}\vec{E}_0\vec{P}. \tag{21}$$

Here g is the constant of interaction between the point unit field and vacuum fields; it is convenient to express this constant in a normalized form: $g = g_1 e$. By varying the Lagrange function over $\vec{P}$, we obtain a non-homogeneous equation for polarization:

$$\left(\sigma - \Delta + \frac{\partial^2}{\partial t^2}\right)\vec{P} = g_1\vec{E}_0(\vec{r},t); \tag{22}$$

Divergence of the left and the right parts of the equation (22) results in the expression for the induced charge density, related to the electrical polarization for the case when a source is moving with velocity $\vec{v}$:

$$\left(\sigma-\Delta+\frac{\partial^2}{\partial t^2}\right)\rho_e(\vec{r},t)=-g_1\delta(\vec{r}-\vec{v}t). \tag{23}$$

It is obvious from (23) that the induced charge density is the Green function for a point source that fulfills the homogeneous equation over the entire space except one point; but due to this point, the function acquires quite definite values over the entire space.

At first, we consider a particular solution of equation (23). Fourier-transformation over coordinates results in the Fourier-harmonics for the induced charge density in vacuum:

$$\rho_{ek}=-\frac{g_1}{\sigma+k^2-\left(\vec{k}\vec{v}\right)^2} \tag{24}$$

Here the corresponding coordinate dependence of the induced charge density for the case, when velocity lies in z-axis, is:

$$\rho(\vec{r},t)=-\frac{g_1}{(2\pi)^3}\iiint\frac{\exp\left(i\left(k_x x+k_y y+k_z(z-vt)\right)\right)}{\sigma+k_x^2+k_y^2+k_z^2\left(1-v^2\right)}dk_x dk_y dk_z$$

Here, it is convenient to proceed to the new integration variables

$$k_x'=k_x;\;\; k_y'=k_y;\;\; k_z'=k_z\sqrt{1-v^2},$$

in addition, to a new coordinate system:

$$x'=x;\;\; y'=y;\;\; z'=\frac{z-vt}{\sqrt{1-v^2}}.$$

After that, the induced charge density expressed in dimensional units transforms to the equation:

$$\rho(\vec{r},t)=\frac{1}{\sqrt{1-v^2}}\rho'(r');\;\; \rho'(r')=-\frac{g}{4\pi\xi_e^2}\frac{1}{r'}\exp\left(-\frac{r'\sqrt{\sigma}}{\xi_e}\right), \tag{25}$$

From (25) we can see that the characteristic dimension of the polarization charge is a definite value equal to the correlation radius or the Compton length of electron: $r_{ñ,e}=\xi_e/\sqrt{\sigma}=c/\omega_{e0}$. The polarization charge moving relative to the absolute coordinate system, in accordance with the Lorentz transformation, is deformed in such a way that its dimensions decrease along the direction of motion

$$r_{ex} = r_{ey} = r_{c,e}; \; r_{ez} = r_{c,e}\sqrt{1-v^2}.$$

Total polarization charge as an integral over the entire space is proportional to the constant of interaction g and the vacuum susceptibility σ^{-1}:

$$q = \int \rho(\vec{r},t)\,d\vec{r} = -\frac{g}{\sigma} \tag{26}$$

The total polarization charge does not depend on the particle velocity that we can interpret as the law of conservation of charge.

We can find a scalar potential for the motionless source from the expression:

$$\Phi(r) = \sum_k \frac{4\pi}{k^2}\rho_k \exp(ikr) = 4\pi g \sum_k \frac{\exp(ikr)}{k^2(\sigma+k^2)} = \frac{g}{\sigma}\frac{1}{r}\left(1-\exp\left(-\frac{r}{r_e}\right)\right) \tag{27}$$

The polarization for the electron is similar to that of the proton within an accuracy of a charge sign

$$\vec{P}_{e,n}(\vec{r}) = \frac{g}{\sigma}\frac{\vec{r}}{r^3}\left[1-\left(1+\frac{r}{r_{c;e,n}}\right)\exp\left(-\frac{r}{r_{c;e,n}}\right)\right]. \tag{28}$$

They only differ in the characteristic wavelength $r_{c,e}$ and $r_{c,n}$. The main feature of the solution for the polarization (28) is an absence of divergence at a point $r = 0$ that leads to the finite value of the particle energy.

Therefore, we can see that the vacuum polarization results in decrease of the source energy by U_0, both electronic and nucleonic modes having the same form:

$$U_{0e,n} = -g\Phi_{e,n}(r=0) = -\frac{g^2}{\sigma r_{c;e,n}} \tag{29}$$

Non-homogeneous equation (23) defines two parameters: the polarization charge q and the radius of a charge localization $r_{c;e,n}$.

In order to determine vacuum parameters, we require that the polarization charge, both for proton and electron, be equal to the electron charge, whereas the particle energy must be equal to the ionization energy of a source out of a potential energy well, which the source creates for itself:

$$q = \frac{g}{\sigma} = e; \tag{30}$$
$$\hbar\omega_{0,e} = \frac{g^2}{\sigma r_{c,e}}; \; \hbar\omega_{0,n} = \frac{g^2}{\sigma r_{c,n}}.$$

By adding the definition of the fine structure constant $\alpha = e^2/\hbar c$ to the latter equations, we obtain the equality $g_1 = \sigma = \alpha^{-1} \approx 137$. It follows that the vacuum polarizability σ^{-1} equals to the fine structure constant α, whereas the constant of the interaction of a point source with vacuum fields equals to the Dirac monopole charge $g = e/\alpha$ [7].

We can define the correlation radius and the fundamental frequency for electronic and nucleonic normal mode as follows:

$$r_{c,e} = \xi_e \sqrt{\alpha}, \qquad \omega_{e,0} = \frac{1}{\tau_e \sqrt{\alpha}};$$

$$r_{c,n} = \xi_n \sqrt{\alpha}, \qquad \omega_{n,0} = \frac{1}{\tau_n \sqrt{\alpha}}.$$

As a result, the rest energy in form (30) reduces to a quite transparent form:

$$\varepsilon_{e,0} = \hbar\omega_{e0} = \frac{eg}{r_{c,e}}; \quad \varepsilon_{n,0} = \hbar\omega_{n0} = \frac{eg}{r_{c,n}} \tag{31}$$

It follows that the point source with the interaction constant g polarizes vacuum and induces charges with dimension $r_{c,e}$ for electronic and $r_{c,n}$ nucleonic modes. The electric field energy for both proton and electron is equal to $e^2/r_{c,e}$ and to $e^2/r_{c,n}$, correspondingly, and it turns out to be 137 times less than the energy related to the electrical polarization. We should notice that the solution for the polarization (28) is formed by three modes of normal vacuum oscillations P_x, P_y, P_z , each creating a charge equal to $e/3$:

$$\frac{1}{4\pi}\int \frac{\partial P_x}{\partial x} d\vec{r} = \frac{1}{4\pi}\int \frac{\partial P_y}{\partial y} d\vec{r} = \frac{1}{4\pi}\int \frac{\partial P_z}{\partial z} d\vec{r} = \frac{e}{3}.$$

Now we consider the structure of the fields in the excited state. The excited state corresponds to the generation of a source at a certain time. Suppose, a point source is generated at time $t = 0$ under the initial conditions for polarization:

$$\vec{P}(\vec{r},t)\big|_{t=0} = 0; \quad \frac{\partial \vec{P}(\vec{r},t)}{\partial t}\bigg|_{t=0} = 0.$$

In this case, the general solution of equation (22) consists of a particular solution (28) and two fundamental solutions of the homogeneous wave equation:

$$\vec{P}(\vec{r},t) = \vec{P}^{(0)}(\vec{r}) + \vec{P}^{(+)}(\vec{r})\exp(i\omega_0 t) + \vec{P}^{(-)}(\vec{r})\exp(-i\omega_0 t) \tag{32}$$

By taking into account initial conditions, we can reduce the solution (32), both for electron and nucleon, to the form:

$$\vec{P}_{e,n}(\vec{r},t) = 2e\frac{\vec{r}}{r^3}\left[1-\left(1+\frac{r}{r_{c;e,n}}\right)\exp\left(-\frac{r}{r_{c;e,n}}\right)\right]\sin^2\left(\omega_{e,n;0}t\right) \qquad (33)$$

The characteristic feature of the solution above is that the electrical polarization for both electron and proton, covers the entire infinite space and oscillates synchronously with the frequency $\omega_{e,n;0}$.

The solution (33), however, contains a substantial disadvantage: such wave packet cannot move in space, it is a typical standing wave. Impossibility of motion is caused by the fact that the phase velocity of different harmonics $v_f = \omega_k / k$ changes from infinity to the light velocity *c*, whereas the group velocity $v_g = \partial\omega_k / \partial k$ changes from zero to *c*.

In a general case, the solution for the polarization for a wave packet moving with velocity $\vec{v}$ should have a soliton form:

$$\vec{P}(\vec{r},t) = \vec{P}(\vec{r}-\vec{v}t)f(t) \qquad (34)$$

A similar property is natural for the solution of a one-dimensional D'Alambert equation that fulfills the condition of deviation from a state of equilibrium for a flexible infinite string $u(x,t) = u(x \pm ct)$. A possibility of motion without changing the form is directly connected to a linear excitation spectrum in k-space $\omega_k = ck$. For two- and three-dimensional cases, the solution of the D'Alambert equation substantially differs from the one-dimensional one. An excitation generated in some point starts spreading (propagating?) at velocity *c* in the form of concentrated circles for two-dimensional case, and in the form of concentrated spheres for the three-dimensional case. The propagation of radio waves strictly follows the three-dimensional D'Alambert equation, which proceeds from the Maxwell equations. Radio waves, however, are a multiquantum process. Nevertheless, a single quantum, while having wave properties, yet behaves like a particle. The thing is that a light quantum radiated by an excited atom at a distant star can cover million years without spreading dispersion. After colliding with a similar atom on the Earth, the light quantum transfers into a similar state of excitation. Therefore, there must be a solution of a soliton type for a light quantum in the form (34), which gives the origin of ray optics.

Analysis shows that it is impossible to obtain such a spectrum in a three-dimensional isotropic space for one order parameter. Following strictly the terminology, we should consider electromagnetic oscillations as coupled oscillations of a two-component order parameter in the form of an electric and magnetic polarization of vacuum.

Suppose, a magnetic polarization with the same Hamiltonian, as that for the electric polarization (7) is possible to appear in vacuum:

$$H = \frac{1}{8\pi}\int\left(\tau_{e,n}^2\dot{\vec{M}}^2 + \xi_{e,n}^2\left(\nabla\vec{M}\right)^2 + \sigma\vec{M}^2\right)d\vec{r} \qquad (35)$$

We define a magnetic order parameter, as well as an electric polarization, through the sum of the elementary magnetic moments:

$$\vec{M} = \frac{4\pi}{\delta V}\sum \vec{\mu}_i.$$

Practice shows that electric and magnetic dipole moments create, correspondingly, electric and magnetic fields, similar in configuration:

$$\vec{E}(\vec{r}) = \frac{3(\vec{r}-\vec{r}')\left(\vec{d}(\vec{r}-\vec{r}')\right)}{|\vec{r}-\vec{r}'|^5} - \frac{\vec{d}}{|\vec{r}-\vec{r}'|^3};$$
$$\vec{H}(\vec{r}) = \frac{3(\vec{r}-\vec{r}')\left(\vec{\mu}(\vec{r}-\vec{r}')\right)}{|\vec{r}-\vec{r}'|^5} - \frac{\vec{\mu}}{|\vec{r}-\vec{r}'|^3}. \quad (36)$$

It follows that for a similar distribution of the electric and magnetic polarization, electric and magnetic fields will be similar as well. We can reduce expressions (36) to the form:

$$\vec{E}(\vec{r}) = -\nabla\Phi_d(\vec{r});\ \ \Phi_d = \vec{d}\nabla_{r'}\Phi_0(\vec{r}-\vec{r}');$$
$$\vec{H}(\vec{r}) = -\nabla\Phi_\mu(\vec{r});\ \ \Phi_\mu = \vec{\mu}\nabla_{r'}\Phi_0(\vec{r}-\vec{r}'), \quad (37)$$

Here Φ_0 is the potential Coulomb function for a unit source $\Phi_0 = 1/|\vec{r}-\vec{r}'|$. Under an arbitrary distribution of the electric and magnetic polarization, scalar potentials (37) acquire the form:

$$\Phi_P = \frac{1}{4\pi}\int \vec{P}(\vec{r}')\nabla_{r'}\Phi_0(\vec{r}-\vec{r}')d\vec{r}' = -\frac{1}{4\pi}\int \frac{div\vec{P}(\vec{r}')}{|\vec{r}-\vec{r}'|}d\vec{r}' = \int \frac{\rho_e(\vec{r}')}{|\vec{r}-\vec{r}'|}d\vec{r}';$$
$$\Phi_M = \frac{1}{4\pi}\int \vec{M}(\vec{r}')\nabla_{r'}\Phi_0(\vec{r}-\vec{r}')d\vec{r}' = -\frac{1}{4\pi}\int \frac{div\vec{M}(\vec{r}')}{|\vec{r}-\vec{r}'|}d\vec{r}' = \int \frac{\rho_\mu(\vec{r}')}{|\vec{r}-\vec{r}'|}d\vec{r}', \quad (38)$$

It follows that the sources of the electric field are the electric charges defined by the relation $div\vec{P} = -4\pi\rho_e$, whereas the sources of the magnetic field are the magnetic charges defined by the relation $div\vec{M} = -4\pi\rho_\mu$. As a result, the electric and magnetic fields meet the conditions:

$$div\vec{E} = 4\pi\rho_e;\ \ div\vec{H} = 4\pi\rho_\mu.$$

Energy of the electric and magnetic fields turns up to be 137 times less than that of the electric and magnetic polarization, correspondingly. The configurations of the electric and magnetic fields are similar under the similar distribution of the electric and magnetic polarization. For example, if we create a homogeneous electric polarization $\vec{P}$ in a full-sphere, then it causes generation of the depolarizing electric field inside the sphere $\vec{E} = -\vec{P}/3$

; therefore, the depolarization coefficient for a sphere is equal to $1/3$. The situation is the same with a spherical magnet: $\vec{H} = -\vec{M}/3$. Generation of the magnetic field also leads to the long-range Coulomb interaction between the normal oscillations: $M_{ex}, M_{ey}, M_{ez}, M_{nx}, M_{ny}, M_{nz}$.

We can define the electric current with the expression:

$$4\pi \vec{j}_e = \dot{\vec{P}} + rot\vec{M}. \; U_0 = -\frac{g}{4\pi}\vec{E}_0\vec{P}.$$

The continuity equation follows from here:

$$div\vec{j}_e + \frac{\partial \rho_e}{\partial t} = 0.$$

Now we show in what way the interaction between the electric and magnetic polarization provides the solution of the soliton type. We add the interaction energy of currents to the Hamiltonians (7) and (35) in the form:

$$U = \frac{\xi_e \tau_e}{4\pi}\int\left(\dot{\vec{P}} \cdot rot\vec{M} - \dot{\vec{M}} \cdot rot\vec{P}\right)d\vec{r} \tag{39}$$

From there we obtain the combined equations for a plane polarized electromagnetic wave:

$$\begin{aligned} &\left(\sigma - \Delta + \frac{\partial^2}{\partial t^2}\right)P_x(\vec{r},t) - 2\frac{\partial}{\partial t}\frac{\partial}{\partial z}M_y(\vec{r},t) = 0; \\ &-2\frac{\partial}{\partial t}\frac{\partial}{\partial z}P_x(\vec{r},t) + \left(\sigma - \Delta + \frac{\partial^2}{\partial t^2}\right)M_y(\vec{r},t) = 0. \end{aligned} \tag{40}$$

By setting up the solutions in the form:

$$\begin{aligned} &P_x(\vec{r},t) = P_x \exp\left(i\left(\vec{k}\vec{r} + \omega t\right)\right); \\ &M_y(\vec{r},t) = M_y \exp\left(i\left(\vec{k}\vec{r} + \omega t\right)\right), \end{aligned} \tag{41}$$

we can obtain the system of equations:

$$\begin{aligned} &\left(\sigma + k^2 - \omega^2\right)P_x - 2\omega k_z M_y = 0; \\ &-2\omega k_z P_x + \left(\sigma + k^2 - \omega^2\right)M_y = 0. \end{aligned} \tag{42}$$

The compatibility condition for the equations (42) leads to the equation:

$$\begin{vmatrix} \sigma + k^2 - \omega^2 & -2\omega k_z \\ -2\omega k_z & \sigma + k^2 - \omega^2 \end{vmatrix} = 0,$$

This gives the spectrum of normal oscillations:

$$\omega = \pm k_z \pm \sqrt{\sigma + k_x^2 + k_y^2 + 2k_z^2},$$

After that, the solutions for the electric and magnetic polarization transmitting with the light velocity reduce to the soliton form:

$$P_x(\vec{r},t) = a \cdot \exp\left(i\left(k_x x + k_y y + k_z (z \pm t)\right)\right) \exp\left(i\left(\pm\sqrt{\sigma + k_x^2 + k_y^2 + 2k_z^2} \cdot t\right)\right);$$
$$M_y(\vec{r},t) = a \cdot \exp\left(i\left(k_x x + k_y y + k_z (z \pm t)\right)\right) \exp\left(i\left(\pm\sqrt{\sigma + k_x^2 + k_y^2 + 2k_z^2} \cdot t\right)\right).$$

We proceed from the supposition that the electron radiates a light quantum; then from a wide range of possible solutions we should choose a solution compatible with the own field of the electron. Since the light quantum propagating along z-axis has a wave vector k_z, we can specify a Fourier-harmonic k_z from the scalar potential (27) which defines the field of the electron. In the cylindrical coordinate system, the Fourier-harmonic for the scalar potential becomes:

$$\Phi_{k_z} = \int_{-\infty}^{\infty} \Phi \exp(-ik_z z)dz =$$
$$= q\left(K_0(rk_z) - K_0\left(r\sqrt{2\sigma + k_z^2}\right)\right).$$

Here K_0 is the Macdonald function. We express the electric polarization along the x-axis as $P_x = \partial\Phi/\partial x$. Thus, for a plane-polarized wave compatible with the field of the electron and fulfilling the system of equations (40), we obtain the solution for the electric polarization:

$$P_x = q\left(k_z K_1(rk_z) - \sqrt{2\sigma + k_z^2} K_1\left(r\sqrt{2\sigma + k_z^2}\right)\right) cos\varphi \exp\left(i(k_z(z-t) \pm \sqrt{2\left(\sigma + k_z^2\right)} \cdot t)\right)$$
(43)

This solution is a quasi-one-dimensional infinite monochromatic wave propagating at the light velocity along the z-axis and interacting with the similar magnetic polarization M_y. In the transversal direction, the monochromatic wave (43) is localized with the dimension equal to the wavelength, since the Macdonald function at big values of argument approximately equals:

$$K_1(x) \approx \sqrt{\frac{\pi}{2x}} \exp(-x).$$

This precisely corresponds to the experiment, as it is impossible to localize a light ray more than the light wavelength.

Therefore, from the values, which we consider as fundamental $e, \hbar, c, m_e, m_p$, we go over to the set of values, which characterize properties of physical vacuum $\alpha, g, \xi_e, \xi_p, \tau_e, \tau_p$ under the additional condition: $\xi_e / \tau_e = \xi_p / \tau_p = c$. In connection with this, we must change the concepts of mass and matter.

Wave equations can only be applied to the material medium having definite dynamic properties, so the idea of physical vacuum means that the entire infinite space is filled with a definite matter. The particles that we observe – electrons, protons, photons – these are excitations of vacuum in the form of wave packets, which are eigenfunctions of the united system of twelve equations. From the point of view of wave mechanics, we can characterize a wave packet with energy, momentum, angular momentum and oscillation amplitude; specifically for the electric polarization, we define the amplitude by the electric charge. For a multi-component order parameter, the form or symmetry of oscillations is important. In this connection, the concept of a particle mass does not have independent meaning. Researchers introduced the values of mass and charge, as well as Planck constant for particles, in different periods of time and so far, they have considered these values as independent ones. As we showed above, charge quantization and existence of Planck constant are the consequences of correlation effects related to discreteness of physical vacuum. Now it makes sense to study the concept of mass for a wave packet. From practice we know that, if we describe the particle oscillation spectrum with the expression: $\omega_k = \sqrt{\omega_0^2 + c^2 k^2}$ then the particle velocity is equal to the group velocity:

$$\vec{v} = \vec{v}_g = \frac{\partial \omega_k}{\partial \vec{k}} = \frac{c^2 \vec{k}}{\sqrt{\omega_0^2 + c^2 \vec{k}^2}} \tag{44}$$

If we express the wave vector $\vec{k}$ from (44) through the group velocity, we obtain the value of frequency in the form:

$$\omega_k = \sqrt{\omega_0^2 + c^2 \vec{k}^2} = \frac{\omega_0}{\sqrt{1 - \frac{v^2}{c^2}}} \tag{45}$$

Multiplying terms of (45) by $\hbar$, we come to the relativistic expression for the particle energy:

$$E(v) = \frac{\hbar \omega_0}{\sqrt{1 - \frac{v^2}{c^2}}} = \frac{E_0}{\sqrt{1 - \frac{v^2}{c^2}}} \approx E_0 + \frac{E_0}{2} \frac{v^2}{c^2} = E_0 + \frac{m v^2}{2} \tag{46}$$

The expression for the particle mass $m = E_0 / c^2$ follows from the latter equation (46), the concept of mass being not necessary if we specify velocity in terms of light velocity.

The examples given below illustrate how to express some known values in terms of vacuum parameters:

De Broglie wavelength:

$$\lambda_{\partial,e} = \frac{\hbar}{m_e v} = \frac{\hbar}{p} = \frac{\hbar}{\hbar k} = \frac{1}{k}; \quad \lambda_{\partial,n} = \frac{\hbar}{m_n v} = \frac{1}{k}.$$

Here we have to consider k, the particle wave vector, as a quantum number independent of vacuum parameters.

Compton wavelength:

$$\lambda_{K,e} = \frac{\hbar}{m_{e,0}c} = \frac{c}{\omega_{0,e}} = r_{c,e} = \xi_e \sqrt{\alpha};$$

$$\lambda_{K,n} = \frac{\hbar}{m_{n,0}c} = \frac{c}{\omega_{0,n}} = r_{c,n} = \xi_n \sqrt{\alpha}$$

Classical radius of electron:

$$r_{0,e} = \frac{e^2}{m_{0,e}c^2} = r_{c,e}\alpha = \xi_e \alpha^{3/2}.$$

Bohr radius:

$$a_B = \frac{\hbar^2}{m_{0,e}e^2} = \frac{r_{c,e}}{\alpha} = \frac{\xi_e}{\sqrt{\alpha}} \tag{47}$$

Bohr energy:

$$E_B = \frac{m_{0,e}e^4}{\hbar^2} = m_{0,e}c^2\alpha^2 = \frac{eg}{r_{c,e}}\alpha^2 \tag{48}$$

By making Bohr energy equal to photon energy,

$E_B = \hbar\omega_{\gamma,B} = \hbar c k_{\gamma,B}$

we obtain γ – quantum wavelength, which corresponds to Bohr energy

$$\lambda_{\gamma,B} = \frac{1}{k_{\gamma,B}} = \frac{\hbar^3 c}{m_{0,e}e^4} = \frac{r_{c,e}}{\alpha^2} = \frac{\xi_e}{\alpha^{3/2}}.$$

We can express Rydberg constant through vacuum parameters:

$$R = \frac{m_{0.e}e^4}{2\hbar^3} = \frac{1}{2}\omega_{0.e}\alpha^2 = \frac{\alpha^{3/2}}{2\tau_e}.$$

It follows from the above expressions that fine structure constant characterizes not only the fine structure of the hydrogen atomic spectrum but the entire lengths hierarchy of the quantum mechanics as well. It is easy to see that characteristic lengths form a geometrical progression:

$$r_{0,e} = \alpha r_{c,e} = \alpha^2 a_B = \alpha^3 \lambda_{\gamma,B}.$$

All the lengths contain neither Planck constant, nor mass, nor charge of electron. In this connection, it makes sense to express the Schrödinger equation through the natural parameters of physical vacuum.

The Hamiltonian for the Schrödinger equation for a hydrogen atom looks like this:

$$H = -\frac{\hbar^2}{2m_{e,0}}\nabla^2 - \frac{e^2}{r}$$

In this expression, we take the fundamental constants $\hbar, m_{e,0}, e$, which specify the characteristic parameters of a hydrogen atom (47-48), as independent; however, as we demonstrated above, none of these constants ought to be taken as a fundamental one.

We can write the Hamiltonian of the electron in the nuclear field of a hydrogen atom in a different form:

$$H = \hbar\sqrt{\omega_{0,e}^2 + c^2k^2} - \frac{e^2}{r} \tag{49}$$

The Planck constant expressed through the electron charge reduces (49) to the form

$$H = \frac{e^2\omega_{0,e}}{\alpha c}\sqrt{1 + \frac{c^2k^2}{\omega_{0,e}^2}} - \frac{e^2}{r} = \frac{eg}{r_{c,e}}\left(\sqrt{1 + r_{c,e}^2k^2} - \frac{\alpha r_{c,e}}{r}\right) \tag{50}$$

Here, it is convenient to use the dimensionless length $r \Rightarrow r/r_{c,e}$, the dimensionless wave vector $\vec{k} \Rightarrow \vec{k} \cdot r_{c,e}$ and the dimensionless frequency $\omega_k = \sqrt{1+k^2}$. We express the energy in terms of the electron rest energy $eg/r_{c,e}$:

$$H = \sqrt{1+k^2} - \frac{\alpha}{r} \approx 1 + \frac{1}{2}k^2 - \frac{\alpha}{r} \tag{51}$$

The particle velocity is equal to the group velocity of the wave packet $\vec{v} = \partial\omega_k / \partial\vec{k} = \vec{k}/\sqrt{1+k^2} \approx \vec{k}$ and we express it in terms of light velocity. Approximate expressions correspond to the case of a low velocity $k \approx v << 1$. We can regard the value k^2 in the approximate expression (51) as the eigenvalue of the Laplacian operator; then we may reduce (51) to the equation for the eigenfunction and the eigenvalue:

$$H\psi = \varepsilon\psi; \qquad H = -\frac{1}{2}\nabla^2 - \frac{\alpha}{r} \tag{52}$$

From (52) it follows that the Schrödinger equation only contains one dimensionless small parameter α of a physical vacuum susceptibility. The fundamental function ψ of a free electron in Cartesian coordinates is equal to $\exp(i\vec{k}\vec{r})$; we express the eigenvalue by the equality: $\varepsilon = k^2/2$.

Now we find out the Bohr quantization conditions for a hydrogen atom. The circular motion of electron around an atomic nucleus is defined by the equality of centrifugal and centripetal forces:

$$r\dot{\varphi}^2 = \frac{\alpha}{r^2}. \tag{53}$$

Bohr assumed a quantization of adiabatic invariants:

$$\oint p_i dq_i = nh$$

For the circular motion, the latter relation reduces to the form:

$$pr = n\hbar$$

Externally, it looks as if a quantum of action existed, that provides quantization of a pulse moment. However, by taking into consideration the pulse $p = \hbar k$, we come to the cyclic boundary conditions for a wave vector:

$$kr = n.$$

It follows that Planck constant has nothing to do with forming the wave function. Since $k = v = r\dot{\varphi}$, we can add to equation (53):

$$r^2\dot{\varphi} = n$$

From where we can obtain the energy, radius and velocity at the stationary Bohr orbits:

$$\varepsilon_n = -\frac{\alpha^2}{2n^2}; \quad r_n = \frac{n^2}{\alpha}; \quad v_n = \frac{\alpha}{n}.$$

That accurately corresponds to the relations (47, 48).

Compton scattering, which we regard as one of the evidences proving existence of quantum of action, proceeds from the laws of conservation of energy and momentum for electron and γ – quantum:

$$m_{e,0}c^2 + \hbar\omega_\gamma = \frac{m_{e,0}c^2}{\sqrt{1-\frac{v^2}{c^2}}} + \hbar\omega'_\gamma;$$

$$\hbar\vec{k}_\gamma = \frac{m_{0,e}\vec{v}}{\sqrt{1-\frac{v^2}{c^2}}} + \hbar\vec{k}'_\gamma. \qquad (54)$$

It follows from (54), that we can specify the wave vector of a scattered light by the relation:

$$k'_\gamma = \frac{k_\gamma}{1+\frac{\hbar k_\gamma}{m_{e,0}c}(1-\cos\theta)}, \qquad (55)$$

Here θ is the angle between vectors $\vec{v}$ and $\vec{k}'$; besides, there is a length parameter $\lambda\!\!\!^{-}_{K,e} = \hbar/m_{e,0}c$ where we take the values $\hbar, m_{e,0}, c$ as the fundamental ones. However, by taking into consideration the fact, that relations (5) and (6) define the spectrum of a particle, we can reduce combined equations (54) to the form:

$$\omega_{e,0} + ck_\gamma = \sqrt{\omega_{e,0}^2 + c^2k^2} + ck'_\gamma;$$
$$\vec{k}_\gamma = \vec{k} + \vec{k}'_\gamma,$$

It follows that the scattering characteristic is defined neither by the Planck constant nor by the electron mass, but by the space and frequency resonance for the wave packets; scattering being submitted to the same formula (55) with the Compton length $\lambda\!\!\!^{-}_{K,e} = c/\omega_{e,0} = r_{c,e} = \xi_e\sqrt{\alpha}$ equal to the correlation radius.

Once in his days Planck supposed that radiation and absorption of light should proceed by quanta. Later this brilliant supposition was confirmed. After that, scientists had only to examine the properties of electron responsible for light radiation and absorption in a quantum way. Albert Einstein, however, considered something different. Since we can observe light quanta, then light is quantized due to existence of quantum of action; the question "Why?" being quite inappropriate here since physical mechanism for quantization of action just does not exist. We can only say that these are the properties of space-time. We just substitute one senseless statement by another one. Nevertheless, proceeded from the fact that electron radiates and absorbs light per quanta, a planetary model of electron is suggested by itself. The electron rest energy equals to: $\varepsilon_{0,e} = \hbar\omega_{0,e} = eg/r_{c,e}$. We can write γ – quantum energy in a similar way: $\varepsilon_\gamma = \hbar\omega_\gamma = egk_\gamma$. Since the photon spin equals to $\hbar$, then, by representing it in the form of the orbital moment $s = pr = \hbar k_\gamma r_\gamma = \hbar$, we come to quite transparent cyclic conditions for the radius of photon orbit $k_\gamma r_\gamma = 1$. After that, the photon energy reduces to the form: $\varepsilon_\gamma = eg/r_\gamma$. We can obtain such an energy as follows: use the solution for the electron polarization in the form (28), set it up into Hamiltonian (7) and integrate over space from infinity to the radius r_γ. Therefore, the nature of fields for photon and electron is the same. By radiating photon, an electron takes off some part of its polarization coat, the intrinsic energy of the electron being reduced.

GRAVITAITIONAL OPTICS

In the previous part we showed that all particles can be considered as excitations of physical vacuum; they are the solutions of the unified system of equations for coupled oscillations of the multicomponent order parameter $(\vec{P},\vec{M})$. That is why we can be sure to a certain degree that all particles similarly contribute to the gravitational interaction, particle energy being the interaction parameter. Now we write down the standardized form of the Hamiltonian for a particle in the gravitational field caused by a massive body of mass m_1

$$H = \varepsilon(\vec{p}) - \frac{\gamma m m_1}{r} \tag{56}$$

We express the particle mass through energy $m = \varepsilon(\vec{p})/c^2$; after that the Hamiltonian (56) reduces to the form:

$$H = \varepsilon(\vec{p})\left(1 - \frac{r_g}{r}\right); \qquad r_g = \frac{\gamma m_1}{c^2}. \tag{57}$$

Here r_g is the gravitational radius which scales gravitational potential of a massive body. For an arbitrary potential, the Hamiltonian has the form:

$$H = \varepsilon(p)(1 - \Phi(\vec{r})).$$

In general case, particle energy is defined by the following expression:

$$\varepsilon(\vec{p}) = \sqrt{\varepsilon_0^2 + c^2\vec{p}^2} \tag{58}$$

In addition, particle velocity equals to

$$\vec{v} = \frac{\partial \varepsilon(\vec{p})}{\partial \vec{p}} \tag{59}$$

From the coordinate system (x, y, z, t) we proceed to a new time $t \Rightarrow ct$ and, in equation (58) – to a new momentum $c\vec{p} \Rightarrow \vec{p}$; then the particle velocity does not depend on the chosen scales of length and time, but becomes a dimensionless value expressed in terms of light velocity:

$$\vec{v} = \frac{\partial \varepsilon(\vec{p})}{\partial \vec{p}} = \frac{\vec{p}}{\sqrt{\varepsilon_0^2 + \vec{p}^2}} \tag{60}$$

The second equation that defines the particle motion in the gravitational field looks like this:

$$\dot{\vec{p}} = -\nabla H = \varepsilon(\vec{p})\nabla\Phi \tag{61}$$

By taking into account equation (60), we can rewrite (61) as follows:

$$\frac{d}{dt}(\vec{v}\varepsilon(p)) = \varepsilon(p)\nabla\Phi(\vec{r}) \tag{62}$$

For low velocities we can substitute value $\varepsilon(p)$ by an approximate expression ε_0; after that equation (62) reduces to that of Newton's mechanics:

$$\ddot{\vec{r}} = \nabla\Phi(\vec{r}).$$

Based upon this equation, Albert Einstein affirmed that the inertial mass and the gravitating mass are equivalent. This statement, however, is incorrect. An accurate equation of motion (62) is transformed to:

$$\ddot{\vec{r}} = \nabla\Phi(\vec{r}) - \vec{v}(\vec{v}\nabla\Phi(\vec{r})) \tag{63}$$

It follows, that particle inertia depends on the direction of motion. It is interesting to note that the intrinsic (internal) energetic properties of a particle are lost in the equation of motion (63). This means that we can apply the obtained equation to any relatively compact object. It can be a planet, a satellite, an electron, a proton, a photon, a neutrino – all the same.

Bearing in mind (60), we reduce (61) as follows:

$$\frac{\vec{p}\dot{\vec{p}}}{\varepsilon_0^2 + p^2} = \vec{v} \cdot \nabla\Phi(\vec{r}),$$

From the latter equation we obtain the integral of motion in two different forms:

$$\varepsilon(p)\exp(-\Phi) = const;$$
$$\left(1-v^2\right)\exp(2\Phi) = const. \tag{64}$$

Now we examine the motion in the Coulomb potential with the Hamiltonian (57). In a centrally symmetrical field motion develops in a plane crossing the centre of a massive body; therefore, we can re-write equation (63) for the plane (x, y):

$$\ddot{x} = -\frac{r_g}{r^3}\left(x - \dot{x}(x\dot{x} + y\dot{y})\right);$$
$$\ddot{y} = -\frac{r_g}{r^3}\left(y - \dot{y}(x\dot{x} + y\dot{y})\right). \tag{65}$$

In the polar coordinate system equations (65) becomes

$$\ddot{r} - r\dot{\varphi}^2 = -\frac{r_g}{r^2}\left(1 - \dot{r}^2\right);$$
$$r\ddot{\varphi} + 2\dot{r}\dot{\varphi} = \frac{r_g}{r}\dot{r}\dot{\varphi}. \tag{66}$$

The second equation in (66) can be integrated easily; after that we obtain the integral of motion corresponding to the momentum conservation law:

$$r^2\dot{\varphi}\exp\left(\frac{r_g}{r}\right) = const. \tag{67}$$

At the beginning, we consider a circular motion: $\ddot{r} = \dot{r} = 0$. Then, the first equation of (66) leads to

$$r\dot{\varphi}^2 = \frac{r_g}{r^2}; \quad \dot{\varphi} = \frac{r_g^{1/2}}{r^{3/2}}, \tag{68}$$

This exactly coincides with the results of Kepler's problem, the first space velocity on the orbit of radius r being equal to

$$v_1 = r\dot{\varphi} = \sqrt{\frac{r_g}{r}} \tag{69}$$

Consequently, the first space velocity attains to the light velocity at $r = r_g$. Further, we consider an arbitrary motion relative to a heavy centre. Let us assume that at time t = 0, a particle has coordinates $(r = r_0, \varphi = 0)$, complete velocity v_0 and azimuth velocity $v_{\varphi 0} = r_0\dot{\varphi}_0$. From the integrals of motion (64, 67) it follows:

$$1-v^2=1-\dot{r}^2-r^2\dot{\varphi}^2=(1-v_0^2)\exp\left(2\left(\frac{r_g}{r_0}-\frac{r_g}{r}\right)\right);$$

$$r^2\dot{\varphi}=r_0^2\dot{\varphi}_0\exp\left(\frac{r_g}{r_0}-\frac{r_g}{r}\right)$$

(70)

From the system of equations (70) we obtain the equation that combines φ and r:

$$d\varphi=\frac{v_{\varphi 0}dr}{r\sqrt{\left(\frac{r}{r_0}\right)^2\left(\exp\left(2\left(\frac{r_g}{r}-\frac{r_g}{r_0}\right)\right)-1+v_0^2\right)-v_{\varphi 0}^2}}$$

(71)

Now proceed to a new variable $\rho=r/r_0$ and new parameters of the problem:

$$\beta=r_g/r_0;\ v_{\varphi 0}=\sqrt{\beta(1+\delta)}.$$

(72)

We examine the situation when the radial velocity at the starting point is zero. It follows that $v_0=v_{\varphi 0}$; after this, equation (71) acquires the form

$$d\varphi=\frac{v_{\varphi 0}dr}{r\sqrt{\left(\frac{r}{r_0}\right)^2\left(\exp\left(2\left(\frac{r_g}{r}-\frac{r_g}{r_0}\right)\right)-1+v_0^2\right)-v_{\varphi 0}^2}}$$

(73)

Here, the parameter δ defines a deviation from the circular motion in an orbit. We apply the equation (73) to the Solar system. The gravitational radius of the Sun equals to 1.5 km. The radius of the terrestrial orbit is 1.5 10^8 km, the radius of Mercury orbit is 0.5 10^8 km, the radius of the solar sphere is 6. 96 10^5 km. The parameter β in (73) is equal to $\beta_E=10^{-8}$ for the Earth planet; to $\beta_M=3\cdot 10^{-8}$ for the Mercury; and to $\beta_S=2{,}1\cdot 10^{-6}$ for the Sun surface. It follows that the circular orbital velocity of the Earth is $v_{\varphi E}=\sqrt{\beta_E}=10^{-4}$. In dimensional terms the circular velocity of the Earth equals to $10^{-4}c=30$ *km/cek* The Mercury moves in an elliptic orbit according to (73), where the value $\beta\ll 1$. Second order expansion in series of the exponent (73) leads to the equation

$$\varphi=\int_1^{\rho}\frac{d\rho\sqrt{1+\delta}}{\rho\sqrt{-(1-\delta-2\beta)\rho^2+2(1-2\beta)\rho-(1+\delta-2\beta)}},$$

(74)

It enables to obtain the orbit path

$$\rho=\frac{1+\delta-2\beta}{1-2\beta+\delta\cos\left(\varphi\sqrt{\frac{1+\delta-2\beta}{1+\delta}}\right)}$$

(75)

We can find the complete revolution of the path from the condition:

$$\varphi\sqrt{\frac{1+\delta-2\beta}{1+\delta}}=2\pi$$

Consequently, the angle gain over one revolution of the path is

$$\varphi\approx 2\pi\left(1+\frac{\beta}{1+\delta}\right)$$

The century displacement of the Mercury perigee means that while the Earth makes 100 revolutions around the Sun, the Mercury makes the number of revolutions equal to $100\left(\beta_M/\beta_E\right)^{3/2}$. From here we obtain

$$\Delta\varphi=100\frac{2\pi\beta_M}{1+\delta_M}\left(\frac{\beta_M}{\beta_E}\right)^{3/2}=21'' \tag{76}$$

The value $\delta_M=0,2$ is the eccentricity of the Mercury elliptic orbit.

From (75) we can obtain the condition when an elliptic orbit transforms into a parabolic path:

$$1-2\beta-\delta=0.$$

It follows that the second space velocity is a little less than that of Kepler's problem and is equal to

$$v_2=v_{\varphi 0}=\sqrt{\beta(1+\delta)}=\sqrt{2\beta(1-\beta)}=v_1\sqrt{2\left(1-v_1^2\right)} \tag{77}$$

Further, we consider the motion of a photon or a neutrino in a gravitational field. In this case, for the equation of motion (71), it is necessary to assume $v_0=v_{\varphi 0}=1$. Then, equation (71) leads to

$$\varphi=\int_1^{\rho}\frac{d\rho}{\rho\sqrt{\rho^2\exp\left(2\beta\left(\frac{1}{\rho}-1\right)\right)-1}} \tag{78}$$

In order to calculate the complete angle of displacement φ_β for a light beam passing a gravitating mass, we move to a new variable $\xi=1/\rho$ and, as a result, we obtain:

$$\varphi_\beta=\int_0^1\frac{2d\xi}{\sqrt{\exp\left(2\beta(\xi-1)\right)-\xi^2}} \tag{79}$$

Integral (79) is divergent at $\beta \to 1$. It proceeds from the fact that at the gravitational radius a photon has a stationary orbit. For $\beta << 1$.

$$\varphi_\beta \approx \pi + 2\arcsin\beta \qquad (79a)$$

it follows that the deviation of a light beam moving, for example, along the Sun surface is $0{,}86''$. The only stationary orbit for a photon is $r = r_g$ that corresponds to the parameter $\beta = 1$. The slightest deviation from unit makes a photon either leave for infinity, or fall down to the centre. Figure 1 illustrates a photon getting off a stationary orbit.

Now we study a radial motion which we can determine from the integrals of motion (68). Under the given input conditions of the coordinate and velocity directed along the radius, and by using the integrals of motion (68), we obtain the energy of a photon moving away from the centre:

$$\varepsilon(p) = p = p_0 \exp\left(-\beta\left(1 - \frac{1}{\rho}\right)\right) \qquad (80)$$

It follows that the photon crosses freely the gravitational radius and at the infinity the photon energy equals to:

$$p = p_0 \exp(-\beta) = p_0 \exp\left(-\frac{r_g}{r_0}\right) \qquad (81)$$

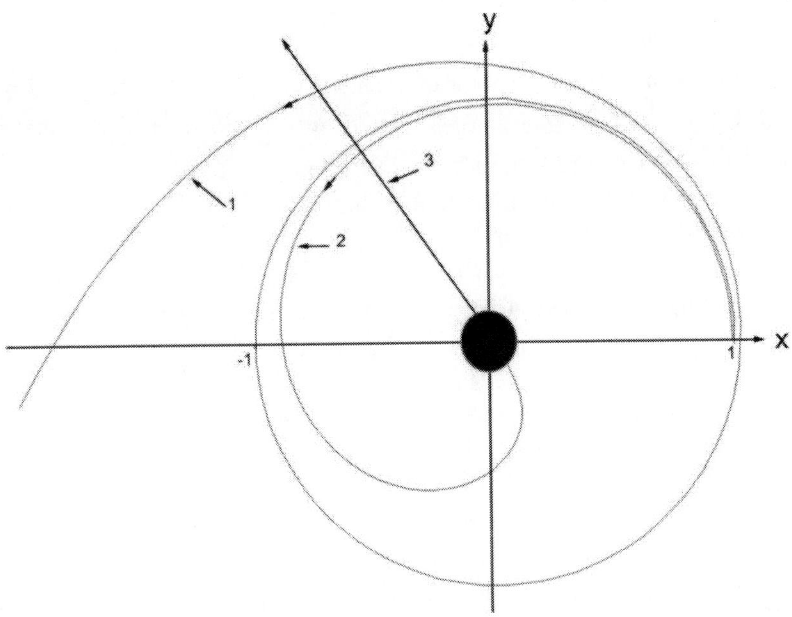

Figure 1. The paths of a photon at different initial conditions. Curve 1 exhibits the photon leaving for infinity at the input condition $\beta = 0{,}9999$. Curve 2 shows the photon falling down to the centre at $\beta = 1{,}01$ Arrow 3 displays the photon radially leaving for infinity from under the gravitational radius.

The radial velocity of the photon remains constant;

$$v_x = \frac{p_x}{p};\; v_y = \frac{p_y}{p};\; v^2 = v_x^2 + v_y^2 = \frac{p_x^2 + p_y^2}{p^2} = 1.$$

The velocity of particles with non-zero mass is defined by the equation:

$$\left(1 - v^2\right)\exp\left(2\frac{r_g}{r}\right) = \left(1 - v_0^2\right)\exp\left(2\frac{r_g}{r_0}\right) \tag{82}$$

From (82) we can define the second space velocity:

$$v_0^2 = v_2^2 = 1 - \exp(-2\beta), \tag{83}$$

it follows from (83) that at the initial velocity $v_0 > v_2$, any particle crosses freely the gravitational radius and leaves for infinity. Note that the first space velocity v_1 equals to $\sqrt{\beta}$. A circular orbit is steady under the condition that $v_2 > v_1$. From the equation

$$\beta = 1 - \exp(-2\beta)$$

we define the boundary of stability for circular orbits $\beta = \beta_c \approx 0,796812$. Circular orbits are only stable to small disturbances under the condition $\beta < \beta_c$. This situation is described by the equation of motion (73) where we can consider the value of δ as a disturbance of a circular orbit. It follows from (73) that for $\beta > \beta_c$ any small value $\delta > 0$ makes a particle leave for infinity along the path similar to that shown on Figure 1 (curve 1). Under the disturbance $\delta < 0$, a particle falls down to the centre and as well leaves for infinity along the curve similar to 2, 3 on Figure 1.

Since all bodies in the Solar system obey the same equation of motion (66), we can measure time in terms of any periodical process that occurs in the Solar system; for example, in terms of revolution of the Earth around the Sun. Further, since we can calculate the periods of revolution for any bodies beforehand, time in the entire Solar system runs similarly. Moreover, we extend the time over the entire visible part of the Universe; and we are quite right when we measure time in billions of years, whereas we measure distance in billions of light years.

Therefore, following Newton, we can repeat that a particle moves uniformly and straight until no force is applied. Following Galilee, we can say that under the same initial conditions in the gravitational field all particles move along the same paths. For example, under the same initial conditions an ultra relativistic proton moves in the same path as a photon does. However, Einstein's statement that time runs differently in each lift does not have any physical meaning, since every electron covers the entire infinite space (33) and simultaneously interacts with all particles in the Universe.

PROBLEMS OF DARK MATTER

In the previous section we introduced the concept that it is the total particle energy $\varepsilon(\vec{p})$ which plays the key part in the gravitational interaction, but not the rest mass, as it is usually considered. This fact substantially changes the estimations of the matter quantity participating in the gravitational interaction. For example, the protons whose energy achieves 10^{21} eV in cosmic rays create a gravitational potential 10^{12} times higher than that for protons on the Earth whose energy is 10^{9} eV. The situation is similar for neutrino. The mean energy of neutrino emitted by neutron beta decay is about 10^{6} eV; whereas zero energy of neutrino, which we usually take into account for gravitational interaction, is estimated by value of 10eV. Consequently, neutrino contributes into the gravitational interaction 10^{5} times more. Photons having zero mass are not considered as carriers of the gravitational interaction at all. Deviation from the straight motion for a photon is caused by the Einstein deflection effect. This point of view contradicts elementary physics. The thing is that, if two bodies exist at positions $\vec{r}_1$ and $\vec{r}_2$, and interact according to the law $U(\vec{r}_1-\vec{r}_2)$, then their momenta $\vec{p}_1$ and $\vec{p}_2$ follow the equations

$$\dot{\vec{p}}_1=-\nabla_{r_1}U(\vec{r}_1-\vec{r}_2);\quad \dot{\vec{p}}_2=-\nabla_{r_2}U(\vec{r}_1-\vec{r}_2). \tag{84}$$

Since

$$\nabla_{r_1}U(\vec{r}_1-\vec{r}_2)=-\nabla_{r_2}U(\vec{r}_1-\vec{r}_2), \tag{85}$$

then, as a consequence of (84 – 85), follows the law of total momentum conservation:

$$\frac{d}{dt}(\vec{p}_1+\vec{p}_2)=0;\quad \vec{p}_1+\vec{p}_2=const. \tag{86}$$

Thus, the distortion of the trajectory for a photon passing e.g. the Sun shows(demonstrates) the variation of its momentum; it follows from the law of the total momentum conservation that the momentum of the Sun changes by the same amount. We can make an obvious conclusion: if a photon is attracted to a massive body, then the massive body is attracted to the photon to the same extent. Therefore, photons, like any other particles, participate in the gravitational interaction, interaction intensity being proportional to the proper intrinsic energy of the particle: $\varepsilon=\hbar\omega=p$.

The azimuth velocity of stars in galaxies is about 100-200 km/sec. That is why, the dark matter elements belonging to a certain galaxy at first sight may seem to have the same velocities. Hence, all relativistic particles, such as photons, neutrino, and cosmic rays, are beyond our consideration; as a result, practically none of the observed particles can create an additional gravitational field. In this connection an idea arises that there are heavy cold particles contributing only to the gravitational interaction; they are called dark matter.

However, a possible alternative point of view exists. First, we examine a simple example. A charged ion of a hydrogen atom creates a Coulomb potential where localized states for an electron are formed. Filling up one of the localized states makes the hydrogen atom electrically neutral, as the nuclear field is completely screened by an electron. On the other hand, if we insert a proton into a metal where there is a sea of free electrons, the localized state does not occur, but this time the nuclear field is screened by free electrons. The trajectory of each electron is distorted near the nucleus so much, that, as a result, electron density increases exactly to the same extent and it screens the nuclear field completely. A positive charge interacts with all free electrons of metal in a Coulomb way and attracts them.

Any heavy body attracts all free particles of a cosmic space by the gravitational interaction in a Coulomb way as well. Nevertheless, there is a significant difference between these two processes. Free electrons of metal are attracted to a positive charge, begin repulsive from each other, as a consequence, the electrical field of the positive charge is screened by electrons. The situation is quite opposite with the gravitational interaction. A massive body attracts particles from the surrounding space. Due to this attraction the total gravitational potential increases, thereby increasing the particle attraction even more. A positive feedback or antiscreening arises that can lead to the system instability. As an illustration, we examine the both situations: screening of an electrical field by free electrons in metal and antiscreening of a gravitational field by free particles (any) in cosmic space.

An external charge with harmonics $\rho_{ext}(k)$ placed into a metal creates a real charge $\rho_i(k)$ defined as a sum of external and induced charges:

$$\rho_i(k) = \rho_{ext}(k) + \rho_{res}(k) \tag{87}$$

We can express the induced charge through the polarizability of electrons in metal $\rho_{res}(k) = -\chi(k)\rho_i(k)$.

Here $\chi(k) = k_{TF}^2/k^2$; k_{TF} is a characteristic wave vector calculated using a Thomas – Fermi approximation [8]. As a result, we obtain

$$\rho_i(k) = \frac{\rho_{ext}(k)}{1+\chi(k)} = \frac{k^2}{k_{TF}^2 + k^2}\rho_{ext}(k);$$
$$k_{TF}^2 = \frac{4k_F}{\pi a_B}, \tag{88}$$

Here k_F is a Fermi momentum in metal. It follows from (88) that a Coulomb potential of a point charge q, for example, is transformed into a screened potential:

$$\Phi_{ext} = \frac{q}{r}; \quad \Phi_i = \frac{q}{r}\exp(-k_{TF}r).$$

Now we consider a situation rather close to the gravitational interaction. Suppose, the entire space is filled up with neutral particles that have some homogeneous density ρ_0 and interact according to the law of gravitation. If any density fluctuation $\rho_{ext}(k)$ occurs in the space, then, owing to the gravitational interaction, all other particles begin to adjust to this density; there-after we can re-write the real density in the form:

$$\rho_i(k) = \rho_{ext}(k) + \rho_{res}(k) = \rho_{ext}(k) + \chi_\gamma(k)\rho_i(k) \quad (89)$$

On the analogy of a free electrons susceptibility, we imagine a gravitational susceptibility $\chi_\gamma(k)$ like this: $\chi_\gamma = k_0^2/k^2$. Here k_0 depends on the value of ρ_0 and on the distribution function of the particle velocity. Afterwards, the real density acquires the form:

$$\rho_i(k) = \frac{k^2}{k^2 - k_0^2}\rho_{ext}(k)$$

This causes gravitational instability of the system relative to the long-wave density fluctuations. As fluctuations develop, slow particles, which compose a small part of an average density, are pulled out of the surrounding space and transformed into clusters of matter in form of stars and galaxies. Fast relativistic particles remain free and continue to participate in creating an additional gravitational field. We denote clusters of a cool matter in form of stars and galaxies having finite motion as $\rho_{cold}(k)$. The remainder relativistic particles in form of cosmic rays, photons and neutrino create additional nonhomogeneous matter density due to the trajectory distortion $\rho_{rel}(k) = \chi_{rel}(k)\rho_i(k)$. Here $\chi_{rel}(k)$ is the gravitational polarizability of the relativistic particles. Thus, the total density is equal to

$$\rho_i(k) = \frac{\rho_{cold}(k)}{1 - \chi_{rel}(k)}.$$

Being on Earth, we have no possibility to scan the distribution of a total energy over the entire space. However, judging from the fact that the azimuth velocity of stars moving away from the centre of galaxy remains nearly constant, the total gravitational potential must have the form:

$$\Phi_i(r) = \eta \ln\left(\frac{r}{R_c}\right),$$

Here η is a dimensionless parameter, R_c – a gravitational size of a space belonging to a certain galaxy. Provided that the centrifugal and centripetal forces are equal

$$r\dot{\varphi}^2 = \frac{\partial \Phi_i(r)}{\partial r} = \frac{\eta}{r}$$

we come to the expression for the circular velocity:

$$v = r\dot{\varphi} = \sqrt{\eta}.$$

At the star velocity being approximately equal to 200 km/sec we obtain the value $\eta = 8 \cdot 10^{-7}$. From the expression for the potential and with the aid of the Poisson equation we obtain the space distribution density of matter:

$$\Delta\Phi_i = \frac{\eta}{r^2} = \frac{4\pi\gamma}{c^2}\rho_i(r);$$

$$\rho_i(r) = \frac{c^2}{4\pi\gamma}\frac{\eta}{r^2}.$$

The space integral of density provides a value of mass inside a sphere of radius R:

$$M_i = \frac{\eta c^2 R}{\gamma} = \frac{v^2 R}{\gamma}.$$

For our Galaxy having the size of about $R = 5 \cdot 10^4$ light years and velocity of $v = 200$ *km/sec* we obtain $M_i = 10^{45}$ *gr*. Mass of cool matter is estimated by value $M_{cold} = 4 \cdot 10^{44}$ *gr.*, therefore, mass of a relativistic matter is comparable with that of the cool one $M_{rel} \approx M_{cold}$. Thus, as a result of the trajectory distortion for relativistic particles, an additional nonhomogeneous distribution of relativistic matter occurs and, consequently, an additional gravitational potential as well. That is why, there is no need to search for a mystical dark matter; relativistic energy is quite sufficient to create an additional gravitational field. Moreover, emission of radiation by stars and galaxies as well as supernova outburst lead to the constant growth of relativistic energy in space. So, observations of the azimuth stellar motion both in galaxies and galaxies in clusters point to the existence of an additional gravitational field. Since azimuth and radial motion follows from the general equation of motion, for example in form (63), the radial motion is submitted to the same additional gravitational attraction; for this reason, there is no dark energy to create antigravitation [9]. Thus, if red shift is related to recession of galaxies, then a contradiction arises, because galaxies have to scatter with acceleration but, judging from the azimuth motion, this is impossible.

It is more natural to consider atomic spectra of far stellar radiation to be time dependent as a consequence of time dependence of physical vacuum parameters. Since atomic levels are proportional to Bohr energy, and Bohr energy, in turn, is proportional to the rest energy of electron ($\varepsilon_B = \alpha^2 \varepsilon_{0e}$) we can affirm that the electron mass increases with time; this means that vacuum parameters for the electron oscillation branch $\tau_e(t)$ and $\xi_e(t)$ decrease with time. The Hubble constant can be defined from the following expression:

$$H(t) = \frac{d}{dt}\left(\frac{m_{e0}(t)}{m_{e0}(t=0)}\right) = \frac{d}{dt}\left(\frac{\tau_e(t=0)}{\tau_e(t)}\right) = \frac{d}{dt}\left(\frac{\omega_{0e}(t)}{\omega_{0e}(t=0)}\right) \qquad (90)$$

The red shift indicates that the Hubble constant is a monotonically growing time function and at the present moment it equals to $2.5 \cdot 10^{-18}$ cek^{-2}.

Nowadays the Universe is in a metastable state, energy emission transitions occurring in two opposite directions. On one hand, nucleosynthesis of light nuclei - takes place, which is the source of stellar energy. On the other hand, nuclear disintegration of heavy nuclei (natural radioactivity) – occurs, as well with energy emission. From today's point of view, nuclear fusion looks quite natural as there is a binding energy between nuclei; moreover, the binding energy on one nucleus increases with the growth of atomic number up to iron. Creation of heavier elements turns out to be less gainful; in this connection it is a surprise that heavy elements, up to uranium, exist on Earth. Nuclei of uranium are in metastable state. If we launched a piece of uranium towards the Sun, the uranium nuclei, under neutron bombardment, would decompose into lighter fragments. This means that uranium cannot occur on Sun. Deposits of uranium on Earth, however, prove that the Earth is an earlier formation than the Sun. Chemical composition of the Earth principally differs from that of the Sun. Sun consists of 75% hydrogen, 24% helium and a negligibly small amount of heavier elements, whereas Earth consists of 32% iron, 30% oxygen and a noticeable amount of heavier elements up to uranium. Heavy elements existing on Earth, as well as the red shift, point out to nonstationarity of physical vacuum parameters. Heavy elements could only occur on Earth when they were energetically gainful; variations of physical vacuum parameters led to the transition of heavy nuclei into a metastable state. A further evidence of nonstationarity of physical vacuum parameters is that not only stars, but also planets emit energy; moreover, volcanic activity, similar to that on the Earth, is still being observed on Jupiter satellites. It is known that the Jupiter emits twice as much energy as it receives from the Sun. We can express the Jupiter energy emission via the Hubble constant. From the law of conservation of energy it follows:

$$\frac{d}{dt}\left(M_J c^2\right) + L_J = 0. \qquad (91)$$

Here $M_J c^2$ – is Jupiter energy of $1.8 \cdot 10^{51}$ erg, and L_J is the integral emission flux of $6.5 \cdot 10^{25}$ erg/sec. By dividing both parts of (91) into Jupiter energy, and considering that the Jupiter only consists of hydrogen, we can reduce (91) to form

$$\frac{d\varepsilon_{0e}/dt + d\varepsilon_{0n}/dt}{\varepsilon_{0e} + \varepsilon_{0n}} + l_J = 0. \qquad (92)$$

Here l_J is specific luminosity of the Jupiter equal to $3.6 \cdot 10^{-26} s\mathring{a}\hat{e}^{-1}$. Taking into consideration the definition of the Hubble constant (90), we can re-write equation (92) as follows:

$$\frac{d\varepsilon_{0n}/dt}{\varepsilon_{0n}} = -\frac{\varepsilon_{0e}}{\varepsilon_{0n}} H(t) - l_J, \qquad (93)$$

All values at the right part of (93) are known, therefore,

$$\frac{d\varepsilon_{0n}}{dt} \approx -\frac{d\varepsilon_{0e}}{dt}.$$

So, the red shift shows that the rest energy of the electron is growing with time, whereas emission of radiation by planets indicates that the rest energy of the proton is decreasing; the total change of the energy for the electron and proton is so great, that it leads to planet heating and emission of radiation.

Since physical vacuum has existed eternally, the values, which characterize the vacuum, can only be of two types: either time independent constants, or oscillating functions. The fundamental values are general for both electron and nuclear modes $g, \alpha, c = \xi_e/\tau_e = \xi_n/\tau_n$ seem to be thought as constant values; however, we have to consider as time dependent the values, which are characteristic either for an electron mode only by ξ_e, τ_e, or for a nuclear one - by ξ_n, τ_n. At the present moment electron and nuclear frequencies are moving towards each other.

Finally, we pay attention to one more mechanism of a gravitational instability. Not coincidentally, there has been some cause for concern so far, that microscopic black holes are possible to occur under the experimental research with Large Hadron Collider in CERN. The thing is that the gravitational attraction between particles grows with increasing particle energy, whereas, the electrical repulsion remains constant due to the law of conservation of charge. In this connection we examine two protons which are speeded up to a certain energy in an accelerator. We write down the Hamiltonian for two protons, taking into account an electrical and gravitational interaction:

$$H = \varepsilon_1 + \varepsilon_2 + \frac{e^2}{|\vec{r}_1 - \vec{r}_2|} - \frac{\gamma m_1 m_2}{|\vec{r}_1 - \vec{r}_2|}.$$

Since masses of the particles are proportional to their energy $m_1 = \varepsilon_1/c^2$, $m_2 = \varepsilon_2/c^2$, , then, under the following condition

$$\frac{\gamma \varepsilon_1 \varepsilon_2}{c^4} > e^2$$

the gravitational attraction turns out to exceed the electrical repulsion. Consequently, the gravitational collapse may occur when the particle energy amounts to

$$\varepsilon > \varepsilon_c = \frac{ec^2}{\sqrt{\gamma}} = 10^{27}\ \text{eV}.$$

Maximum particle energy in cosmic rays reaches 10^{21} eV. The value of energy expected at the accelerator in CERN is $7 \cdot 10^{15}$ eV that is eleven orders less than the critical value. That is why the microscopic black holes are impossible to appear in the accelerator. From the expression for the critical energy, we can define the specific wave vector and the corresponding de Broglie wave length:

$$\varepsilon_c = c\hbar k_c = \frac{c\hbar}{\lambda_c} = \frac{ec^2}{\gamma^{1/2}}$$

It follows:

$$\lambda_c = \frac{e\sqrt{\gamma}}{\alpha c^2} = \sqrt{\frac{\gamma\hbar}{\alpha c^3}} = 1.88 \cdot 10^{-32} cm$$

Planck introduced the specific length by reason of dimension:

$$r_{Pl} = \sqrt{\frac{\gamma\hbar}{c^3}}.$$

It is easy to see that the specific length λ_c can be expressed via Planck length as follows:

$$\lambda_c = \frac{r_{Pl}}{\sqrt{\alpha}}$$

Thus, the considerations above allow attaching a physical sense to the Planck length, which defines the most probable value for the lattice constant of physical vacuum; here k_c specifies the edge of the Brillouin zone in $k-$space, and ε_c – the width of the allowed energy region.

CONCLUSION

In this chapter we try to consider problems of the gravitational optics and dark matter developing from the crystal model for the vacuum. Thus, our model for vacuum is represented as a material medium in which dynamical properties of the crystal specify the spectrum of elementary particles . How it is follows from consideration it enables to describe both electromagnetic waves and spectrum of elementary particles from the unified point of view. We have obtained the combined equations for a multicomponent order parameter in the form of the electric and magnetic vacuum polarization, which defines the spectrum and symmetry of normal oscillations in the form of elementary particles. We have restored the fundamental parameters of physical vacuum, such as: a susceptibility for the electric and magnetic polarization (equal to the constant of fine structure), parameters of length and time for the electron and nuclear branches of the oscillations, correspondingly. We have shown that the charge quantization is directly connected to discreteness of vacuum consisting of particles with the interaction constant equal to the double charge of a Dirac monopole. Elementary particles are excitations of vacuum in a form of wave packets of a soliton type. We have obtained an exact equation of motion for a particle in a gravitational field. Energy defines both gravitational interaction and particle inertia, inertia being of an anisotropic value; that is why the statement, that the inertial and gravitational masses are equivalent, is not correct. We have examined the situation when galaxies are distributed over the entire infinite space according to the cosmological principle. In this case recession of galaxies is impossible; therefore, the red shift of radiation emitted by far galaxies must be interpreted as the blue time shift of atomic spectra. As a consequence, it follows that both rest energy and mass of electron are increasing now. Since physical vacuum exists eternally, vacuum parameters can be either constant or oscillating with time. These are time oscillations of $\xi_e(t), \xi_n(t)$ and $\tau_e(t), \tau_n(t)$ which have caused electron mass growth within recent 15 milliard years, inducing red shift; on the contrary, proton mass decreases, responsible for emission of radiation by planets.

REFERENCES

[1] Zwicky, F., Helv. (1933). *Phys. Acta* 6 110.
[2] Freeman, K. C. (1970). *Astrophys. J.*160 811.
[3] Tyson, J. A. et al. (1984). *Astrophys. J.* 281 L59.
[4] Riess, A. G. et al. (1998). *Astron.J.* 116 1009.
[5] Perlmutter, S at al. (1999). *Astrophys. J.* 517 565.
[6] Milne, E. Q. J. (1934). *Math.* 5 64.
[7] Dirac, P. A. M., Proc. Roy. (1931). *Soc. A.*133, 60.
[8] Pines, D. (1963). *Elementary Excitations in Solid*, New-York – Amsterdam.
[9] Chernin, A. D. (2001). *Phys*, Usp. 44 1099.

In: New Developments in Magnetohydrodynamics … ISBN: 978-1-61942-078-6
Editors: Stephen Talton and Patrick F. Theba 2012 Nova Science Publishers, Inc.

Chapter 2

ORIGIN OF THE SATURN RINGS: ELECTROMAGNETIC MODEL OF THE SOMBRERO RINGS FORMATION

Vladimir V. Tchernyi (cherny)*
SAIBR. 20-2-702, Osennii blvd. Moscow, Russia

ABSTRACT

For the first time the role of superconductivity of the space objects within the Solar system located behind a belt of asteroids is considered. Observation of experimental data for the Saturn's rings shows that the rings particles may have superconductivity. Theoretical electromagnetic modeling demonstrates that superconductivity can be the physical reason of the origin of the rings of Saturn from the frozen particles of the protoplanetary cloud. The rings appear during some time after magnetic field of planet appears. It happened as a result of interaction of the superconducting iced particles of the protoplanetary cloud with the nonuniform magnetic field of Saturn. Finally, all the Kepler's orbits of the superconducting particles are localizing as a sombrero disk of rings in the magnetic equator plane, where the energy of particles in the magnetic field of Saturn has a minimum value. Electromagnetism and superconductivity allow us to understand why planetary rings in the solar system appear only for the planet with the magnetic field after the belt of asteroids where the temperature is low enough and why there are no rings for the Earth, and many other phenomena.

1. INTRODUCTION

There is no yet clear picture of the origin of Saturn's rings. There are two versions of the rings origin. The most common version saying that the rings originated when an asteroid type body approached Saturn and was destroyed by the gravity and centrifugal forces and then from the debris the rings were created. It looks like this version contains a mysterious fact. Another idea relates to the origin of rings from the particles of the protoplanetary cloud around Saturn, and this problem has not been resolved yet. This chapter compensates the lack

* Email address: +7(926)592-6066, chernyV@bk.ru

of this knowledge, demonstrating how rings could originate and form from the frozen particles of the protoplanetary cloud after the appearance of the magnetic field of Saturn due to electromagnetic interaction of icy particles with the planetary magnetic field.

The founder of the theory of electromagnetic waves J.K. Maxwell in his award winning paper on the subject "On the stability of the motion of Saturn's rings" (1859), deduces that the rings of Saturn cannot be solid and the rings could be stable only if they consist of "an indefinite number of unconnected particles orbiting Saturn in much the same way as our Moon orbits the Earth" [1]. Otherwise gravitational forces would destroy them. Ground-based experiments and the data of the Pioneer, Voyager 1, 2 and Cassini-Huygens space missions have revealed the rings to be composed of icy particles, and icy particles with impurities.

After G. Galileo (1610) many researchers have studied the nature of the rings [for example, 1-16]. From the consideration of the gravity and celestial movements it follows that different ring systems are morphologically quite distinct and are all shaped by a few common processes. This is the outward transport of angular momentum by rings particles and by gravitational interactions between satellites, moons and ring material. Orbital resonances between satellites, moons and ring particles play an important role in enhancing the influence of satellites and forming specific structure of the rings and gaps. At the same time, extensive experimental data confirmed the importance of magnetohydrodynamic plasma phenomena and, particularly, electromagnetism of the rings structure origin.

Despite the rich available database, there is no yet physically satisfactory model of the Saturn rings and the mystery of many experimental data has no explanation: origin, evolution and dynamics of the rings; why particles are separated and at the same time they could stick together; considerable flattening and the sharp edges of the rings; thin periodic structure of the rings; deformation of the magnetic field lines nearby ring *F*; formation of "spokes" in the ring *B*; high radio-wave reflectivity and low brightness of the rings; anomalous reflection of circularly polarized microwaves (like from magnetic mirror); strong pulse electromagnetic radiation of the rings in the 20.4 kHz - 40.2 MHz range; spectral anomalies of the thermal radiation of the rings; why substance of the rings does not mix, but preserves its small-scale color differences; existence of an atmosphere of unknown origin nearby the rings; why there is existence of the waves of density and bending waves within the rings; why the planetary rings in the solar system appear for the planets which are located only outside the belt of asteroids and why the Earth has no rings, etc.

The superconductivity of the ring particles may follow from the fact that the ring particles are relics of the early days of the Solar system and particles were never subject to coalescence and heating. Indeed, the Sun heats the rings weakly, because temperature in the area of the rings is about 70-110 K. It makes possible the existence of the superconducting substance in the space behind the belt of asteroids. The superconductive particles cannot stick together because the magnetic field emanates from them and pushes the particles apart. In 1933 W. Meissner and R. Ochsenfeld found that a superconducting material will repel a magnetic field. The high-temperature superconductivity was discovered by Bednortz and Muller in 1986 [17]. In 1986 superconductivity of ice was experimentally demonstrated by A.N. Babushkin[1] et al [18].

[1] Professor Aleksey N. Babushkin is a Dean of the Physics Department at the Ural State University, Yekaterinburg, Russia

It is assumed that superconducting matter of the particles of the rings of Saturn allows extending classical theories of the planetary rings (gravitational, mechanical, magnetohydrodynamic and plasma interactions) by non-conflicting superconducting model [20-44].

An interesting fact of this chapter is that even though its subject concerns problem related to astronomy, the solution may come from the general electromagnetic theory.

2. An Experimental Data Observation

Thin Structure and Sharp Edges of the Rings Similar to magnetic particles creating dense and rarefied areas in a nonuniform magnetic field, the superconductive ice particles also form their groups, which from outside look like a system of rings. Superconducting particles will collapse into a system of rings as the result of their replacement in the area with the less density of the magnetic flow, within plane of magnetic equator, with the force: F = - mdH/dz, where m – magnetic moment of a particle, dH/dz – gradient of intensity of a magnetic field along an axis z of the magnetic dipole. The force of diamagnetic push-out is forming the sharp edges of the ring: F= -mdH/dy, where dH/dy - gradient of intensity of a magnetic field along the radius of the ring. The casual break in the ring will be stabilized by the force of diamagnetic push-out F = - mdH/dx, where dH/dx - gradient of intensity of a magnetic field in tangential direction. Figure 1 shows deformation of the magnetic field lines measured for the ring *F* by the Pioneer mission [5, 6, 12], like it is happened for the superconducting disc in laboratory [17] (Figure 1).

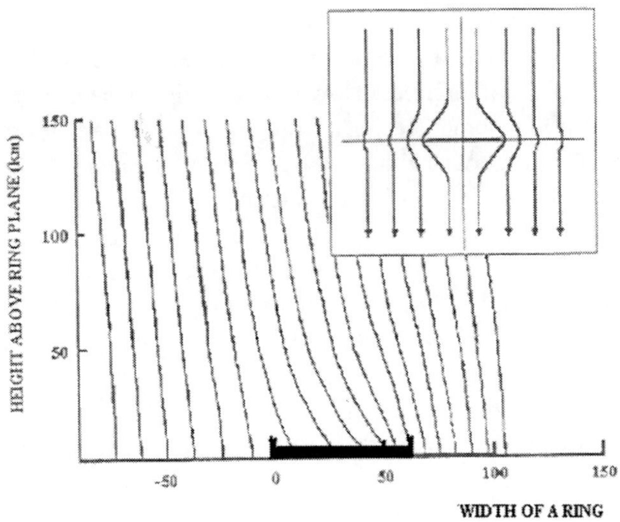

Figure 1. The structure of the magnetic field next to the ring *F* of Saturn [5, 6, 12]. The field of magnetosphere dipole is distorted. The particles flow is located between 0 and 60 km and is directed to the page. The field of dipole is directed downward in the parallel to the negative direction of the negative axis. On the insert picture is the demonstration of the distortion of the magnetic field lines of the superconducting disk under Meissner - Ochsenfeld state [17].

Planetcentrical dust flow. For superconducting particles there is London's depth λ_L of penetration of the magnetic field inside superconductor [17]. Influence of penetration of the magnetic flow becomes appreciable for particles with the size comparable with London's depth of penetration. Smaller particles are not cooperating with the planetary magnetic field, because they lost their superconductivity by size. Dynamics of these particles is different from dynamics of particles with bigger size which is $>2\lambda_L$. These particles will fall down to the planet due to the gravity. Thus, existence of the planetcentrical dust flows of submicron's size particles related to disappearance of superconductivity of the matter of the rings particles due to reducing their size. It is also possible for the particles to loose their superconductivity by influence of collisions and by fluctuations of magnetic field.

Change of the Azimuth Brightness of the A Ring of Saturn There is a number of theories explaining this fact, based on assumptions of a synchronous rotation of the ring's particles with their asymmetrical form as extended ellipsoids directed under a small angle to the orbit, or with asymmetrical albedo of the surface [5, 6, 12]. Let's go to superconducting model. If superconductor is placed in the magnetic field, additional moment directed opposite to the external field arises. The matter is magnetized, not along the external magnetic field but in the opposite direction. The rod of superdiamagnetic substance of the ring particle tries to locate itself perpendicularly to the magnetic field lines. From the science that studies ice [19] it is a known fact that at the temperature below – 22 ^{0}C growing snowflakes take the form of prisms. Thus, the prism of the superconducting iced particle will be oriented perpendicularly to the field lines of the polhoidal and toroidal constituents of the magnetic fields of Saturn. So, it's clear that variable azimuth brightness of the Saturn's rings system *A* relates to orientation of the elongated ellipsoid of superconducting particles to the normal direction to the magnetic field of the planet.

Spokes in the ring B of Saturn, as well as the spokes of any wheel, are located almost along radiuses. According to the laws of Kepler any radial formation should be distorted and washed out for a few tens of minutes. However, experimental data show that lifetime of separated spoke 10^3 - 10^4sec though Kepler washing outs in them are nevertheless proven. The size of spokes themselves is about 10^4km along the radius and about 10^3km along the orbit of the ring. The matter of the spokes consists of micron and submicron size particles [11]. There were many attempts to explain the nature of the spokes. Mostly all theories are based on the action of the force of gravity. At the same time, there were some ideas that registration of rotating spokes somehow related to electromagnetic interaction because of its rotation synchronously going along with the magnetosphere of the Saturn [5, 6, 12]. The analysis of spectral radiant power of spokes provides specific periodicity about 640.6±3.5 min, which is almost coincident to the period of rotation of the magnetic field of Saturn, which is 639.4 min. Moreover, the strong correlation of maxima and minima of activity of spokes with the spectral magnetic longitudes is connected to presence or absence of the radiation of Saturn's Kilometric Radiation (SKR). It confirms the assumption of the dependence of the spokes dynamics on the magnetic field of Saturn and testifies to the presence of large-scale anomalies in the magnetic field of Saturn. Superconducting iced particles of the rings matter are rotating in accordance with Kepler's law, and at the same time magnetic field is rotating along with the planet and has anomalies itself. Superconducting particles coming to anomalies of the magnetic field positions and the balance of the three forces acting with each particle will change due to change of the electromagnetic force. Then particles will try to get another position and observer can see all these movements of the particles on the picture of the rings as a spokes.

High reflection and low brightness of the rings particles in the radiofrequency range we can explain also using superconducting model. The discovery of strong radar-tracking reflection from the rings of Saturn in 1973 was surprising [5, 6, 12]. It turned out that the rings of Saturn actually have the greatest radar-tracking section among all bodies of the Solar system. It was explained by the metallic nature of the particles. The data of the Voyager excludes this possibility. The disk of superconducting particles completely reflects radiation with frequencies below 10^{11}Hz and poorly reflects radiation with higher frequencies, as in the case of superconductor. The superconductors have practically no resistance up to frequencies of 100MHz. At frequencies about 100GHz there comes a limit, above which the frequent quantum phenomena cause a fast increase of resistance, as it is shown on Figure2. Hence a specific picture of the dependence of brightness.

Own Wide Band Pulse Radiation of the Rings. Data from the Voyager have shown that the rings radiate own wide band pulse radiation within the 20 KHz-40,2MHz [5, 6, 12]. These waves, probably, are a result of interaction of charged particles with the particles of ice or destruction and friction of iced particles when co-striking occur. These incidental radio discharges are named as Saturn's Electrostatic Discharges (SED). The average period SED is well determined and was established as 10 hour 10±5 min and 10 hour 11±5 min by Voyager-1, -2.

If the ring has a source of SED, the area of this source can be located at the distance of 107,990 – 109,000km from the planet according to measured periodicity. Data of the experiments can help specify electrodynamics coupling between the planetary ring system and the magnetosphere, in which SKR, SED and activity of spokes are subordinated to longitude regulation. In accordance with superconducting model, approaching of superconducting rings' particles up to distance of about 10^{-8}m, or, the existence of narrowing or dot contact will result in the formation of a weak link (superconducting transition) through which superconducting electrons can be tunneled.

When the difference of phases between superconductors under action of the electrical or magnetic field occurs, the weak link will generate electromagnetic radiation with frequency proportional to power failure on this transition (nonstationary Josephson phenomenon) [17, 20-36]. The radiation frequency is proportional to the voltage in the transition, ν =2eV/h, where 2e/h= 483,6 MHz/μV, e is a charge of electron, h is Plank constant.

Frequency anomalies of thermal radiation of the rings in the 100μm - 1cm range. The measured brightness temperature on the short waves is less than true brightness temperature of the rings, and on the longer waves the rings look much colder than in the case when the radiation corresponds to their physical temperature [5, 6, 12].

On the range 100μm - 1mm brightness temperature of the ring (Figure 2) sharply falls to the meanings smaller than those ones characteristic of an absolutely black body. On the wavelengths longer than 1cm a ring behaves as the diffusion screen, reflecting planetary and cold space radiation. The central part of the spectral range 100μm – 1cm is the most sensitive to the parameter of refraction, and may contain the important determining information of fundamental properties of the substance.

In accordance with the model, under the superconducting condition the electrons do not interact with a crystal lattice and do not exchange energy with it, therefore they cannot transfer heat from one part of the body into another. Hence, when the substance passes into a superconducting condition, its heat conductivity is lowered. This effect can be obvious under temperatures much less critical, when there are very few conventional electrons capable of transferring heat [20-36].

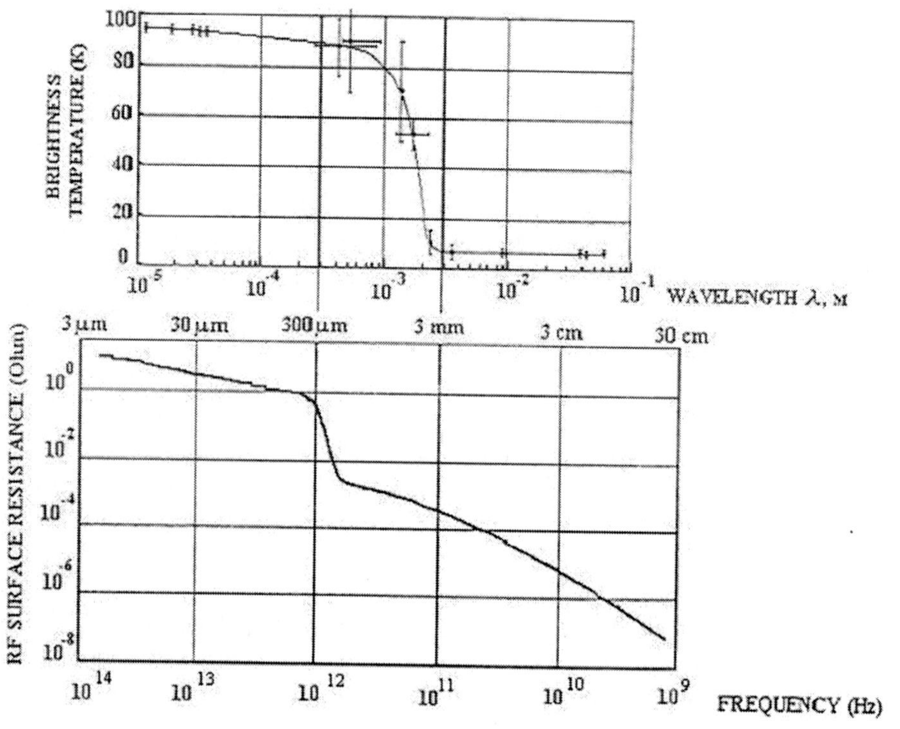

Figure 2. Top diagram is the dependence of the brightness temperature of the rings on the wavelength: transition from the radiation of the almost black body to practically complete reflection is observed [5, 6, 12]. Bottom diagram is the dependence of the surface resistance of the superconductor on frequency for Nb at T=4,2K [17].

Color difference of rings in a small scale. The balance of three forces determines the position of the superconducting particles in the gravitational and magnetic planetary fields: gravitational force, centrifugal one and magnetic levitation (diamagnetic push out), Figure1. Going along with our model let's consider the distribution of three particles (a, b, c) with equal weights on close orbits. Let a particle *a* be wholly superconducting, *b* – have an impurity clathrate-hydrates of ammonia or methane (NH_3; CH_4 H_2O), *c* - has an impurity of sulphur and Ferro containing silicates (H_2S). Each impurity will give the contribution to reduction of the volume of superconducting phase and will determine the color of the particle. Force of diamagnetic push out - F_L depends on the volume of the superconducting phase, therefore for each of considered particles the balance of three forces will be carried out in the orbits with different radiuses [20-36].

Anomalous Inversion Reflection of Microwaves With Circular Polarization Above 1 Cm. The research of reflection of radiowaves above 1 cm from the rings was carried out with the use of ground based radio-locators and by the space probes [5, 6, 12]. The reflection appeared rather large, and the geometrical albedo is equal approximately to 0.34 and has no strong functional dependence on the wavelength or on the angle of the inclination of ring's pitch. The rings are strong depolarizers. Therefore, in order to get any information from the reflection it is necessary to measure separately the intensity of two orthogonal polarized reflected signals. It provides information of the factor of the ring's polarization, which carries information about properties of particles. For the majority of single objects of the Solar system, for example the planets, factor of reflection unobserved polarization (orthogonal to observable) is rather small. As to the rings, the supervision in some range of wavelengths and angles of inclination give reflection factor of unobserved polarization between 0.4 – 1.0 . Let's go to our superconducting model [20-36]. The superconductors have an essential difference from ideal conductors; besides almost infinite conductivity they also demonstrate an ideal diamagnetism. The falling electromagnetic wave will induce circular currents in superconductor, which will completely compensate action of the magnetic field of the incident wave. So that the absence of the magnetic field in the volume of the superconductor should be carried out. Superconductor will be acting as a magnetic mirror. Thus, if the falling on superconductor electromagnetic wave has a determined direction of a circular polarization (the spirality), the direction of circular polarization (spirality) will be kept in the reflected wave.

An Atmosphere of Unknown Origin at the Rings. The atmosphere of Saturn's rings can exist as a result of thin balance of forces of gravitational attraction and diamagnetic push-out of gas molecules. The levitation of gas molecules comes true at the expense of forces of diamagnetic push-out induced in superconducting particles by molecular magnetic moment of gas [20-36]. The similar situation can be observed under laboratory conditions when atmospheric water steam is precipitated on substance as white-frost at the transition moment of substance from a superconducting into a conventional one.

Existence of Wave of Density and Bending Waves Within the Rings. The existence of waves of density and bending waves in the Saturn's rings has no complete explanation based only on gravitation phenomena. Let's use the superconducting model. It is possible to note that the external magnetic field is directed along the free surface of diamagnetic liquid which represents a disk of the rings. In case of periodic deformation of a free surface a normal field under a hollow decreased, and it increased under chamber. Consequently, the ponderomotive force works on the side of restoration of the flat form of the free surface. Thus, the field increases a rigidity of a free surface. The change of rigidity of the free surface in a field gives a chance to excite its parametrical fluctuations. When the quantity of the amplitude of intensity of a variable field is larger than the critical one, the standing wave is occurring on a free surface of superconducting liquid. The constant phase lines of this wave are transverse to the vector of intensity of the field. The wavelength is determined by the condition of the parametric resonance. The constant phase lines begin to be bent when intensity of the field increases, and the excited ripple becomes casual [20-36].

3. Theoretical Solution of the Saturn's Rings Origin

Following from the solution of the electromagnetic problem we will demonstrate how rings of Saturn could be originated from the iced particles located within the protoplanetary cloud. Before appearance of the magnetic field of Saturn all particles within the protoplanetary cloud are located on such an orbit as Kepler's, where there is a balance of the force of gravity and the centrifugal force. With the occurrence of the magnetic field of the

Saturn the superconducting particles of the protoplanetary cloud begin to demonstrate an ideal diamagnetism (Meissner-Ochsenfeld phenomenon). Particles start to interact with the magnetic field and all particles become to be involved in additional azimuth-orbital movement. Let's estimate the result of this movement [37-38].

If the magnetic field of the planet is equal H, and the planetary magnetic moment is equal $\vec{\mu}$, then the magnetic field at any particular point within the protoplanetary cloud, located on the distance $\vec{r}$, can be presented as:

$$\mathbf{H} = \frac{3 \cdot \mathbf{r} \cdot (\mathbf{r}, \boldsymbol{\mu})}{r^5} - \frac{\boldsymbol{\mu}}{r^3} \qquad (3.1)$$

Then the superconducting ball with the radius R, which is located within the protoplanetary cloud, gets the magnetic moment equal:

$$\mathbf{M} = -R^3 \cdot \mathbf{H}. \qquad (3.2)$$

That energy of a superconductor in a magnetic field gets the following value:

$$U_H = -(\mathbf{M}, \mathbf{H}) = R^3 \cdot H^2 \qquad (3.3)$$

If the beginning of coordinates to place in the center of a planet, and an axis z to direct along the magnetic moment of a planet (orthogonal to equator) magnetic energy thus will be equal:

$$U_H = \frac{R^3 \mu^2}{r^6} (3\cos^2\theta + 1) \qquad (3.4)$$

Here θ - an angle between a vector $\vec{r}$ and an axis Z. We can see from the expression (3.4), that magnetic energy of the superconducting particle has a minimum value when the radius-vector $\vec{r}$ (position of the superconducting particle) appears in a plane of magnetic equator, a perpendicular to the axis Z $(\cos\theta = 0)$. It is clear, that for one particle its trajectory (orbit) in the azimuth-orbital movement will be only disturbed by the magnetic field. However, in case of a huge amount of particles, like its happened within the protoplanetary cloud, after some time, collisions between particles will compensate their azimuth-orbital movements, and, as a result, eventually, during some time, all orbits of the particles of the protoplanetary cloud should come together to magnetic equator plane and create highly flattening disc around planet. Within the disc of rings all particles will be located on such an orbit as Kepler's, where there is a balance of the force of gravity, the centrifugal and electromagnetic forces. At the same time, orbital resonances (due to gravity force) between satellites, moons and the rings particles play an important role in forming specific structure of the rings and gaps.

4. Separation and Collision of the Particles Within the Sombrero of Rings

We can define the energy of the interaction of two superconducting particles with magnetic moment μ_1 и μ_2 if they are located on the distance r_1 и r_2, respectively as:

$$U = -\boldsymbol{\mu}_1 \mathbf{H}_2, \quad (4.1)$$

where magnetic field H_2 is produced by the magnetic moment μ_2 it can be presented as

$$\mathbf{H}_2 = \frac{3(\mathbf{r}_1 - \mathbf{r}_2)(\boldsymbol{\mu}_2(\mathbf{r}_1 - \mathbf{r}_2))}{|\mathbf{r}_1 - \mathbf{r}_2|^5} - \frac{\boldsymbol{\mu}_2}{|\mathbf{r}_1 - \mathbf{r}_2|^3} \quad (4.2)$$

If we place the particles with magnetic moment μ_2 at the beginning of coordinate (r_2=0) then the expression for the energy of the interaction of two particles (4.1) will be the following:

$$U = -\frac{3(\boldsymbol{\mu}_1 \mathbf{r}_1)(\boldsymbol{\mu}_2 \mathbf{r}_1)}{|\mathbf{r}_1|^5} + \frac{\boldsymbol{\mu}_1 \boldsymbol{\mu}_2}{|\mathbf{r}_1|^3} \quad (4.3)$$

In the plane of the rings of Saturn the magnetic field of the planet coincides with the rotation axis of the planet. If the axis Z is directed along with the rotation axis of the planet then magnetic moment of the particles also will be directed along the axis Z. Using the cylindrical system of coordinate (ρ, φ, z) we can represent expression (4.3) as:

$$U = -\left(\frac{3z^2}{\left(\rho^2 + z^2\right)^{5/2}} - \frac{1}{\left(\rho^2 + z^2\right)^{3/2}} \right) \mu_{1z}\mu_{2z} = \frac{\rho^2 - 2z^2}{\left(\rho^2 + z^2\right)^{5/2}} \mu_{1z}\mu_{2z}, \quad (4.4)$$

Let us use the expression (4.4) to estimate how two superconducting particles will be interacting in two different cases. The first one is when two particles are located in the same plane within the sombrero of the rings (Z=0), and the second situation is when two particles are located on the different planes but on the same axis $(\rho = 0)$.

From the expression (4.4) follows that for the particles with magnetic moment μ_{1z} and μ_{2z} which are located on the same plane, Z=0, we can get that energy of their interaction is equal:

$$U = \frac{\mu_{1z}\mu_{2z}}{\rho^3} \quad (4.5)$$

From the expression (4.5) follows that in this situation both particles will be pushing each other and they will be holding separation distance in between them. This result has been confirmed by the data of Cassini mission: the particles are separated[2].

Then for another situation both particles are located on the same axis but on the different planes, and, as it follows from (4.4) the expression for the interaction energy is:

$$U = -\frac{\mu_{1z}\mu_{2z}}{|z|^3} \qquad (4.6)$$

We can see that in this case both particles will be attracting each other, they could even collide or stick together and form bigger pieces or lumps of ice. This fact has an experimental conformation by Cassini mission. As we know from the data of Cassini mission it was registered that particles within the sombrero of the rings can collide or even stick together and form bigger pieces of ice[3]. Then later the particles with 50 meters or more in diameter can be destroyed into smaller pieces by the common action of gravity and centrifugal force.

CONCLUSION

Theoretical electromagnetic model of the origin of the rings of Saturn from superconducting particles of the protoplanetary cloud, which is presented in this chapter, is a direct continuation of the paper by J.K. Maxwell (1859). Unfortunately, at his time there was no knowledge about superconductivity (1911) and force of diamagnetic push-out of superconductor (1933). Superconductivity of ice and high temperature superconductivity was discovered just recently, in 1986 [17, 18]. As it follows from above consideration, observation of experimental data and electromagnetic modeling confirmed the suggested model.

The model considered here makes possible a "magnetic coupling" between protosun and superconducting particles of its protoplanetary cloud, that in the process of formation of the solar system at an early stage of its development, when the temperature was low enough to have superconductivity, lead to the carrying of the moment of momentum from the Sun to other planets by the electromagnetic means of the superconducting substance of the space environment. Following this we can conclude that the idea of H. Alfen [10-11] that "solar system history as recorded in the Saturn rings structure" becomes to be a physical reality.

From the approach presented above, with the possibility of an electromagnetic origin for the rings, it follows as important for the space physics to take into account the natural space superconductivity for the space substance after the belt of asteroids. It may have a fundamental importance for analyzing the data of the Cassini-Huygens probe, and striking parallels to those that occur in more remote disc systems such as galactic discs and accretion discs around stars and black holes. The force of diamagnetic push-out of superconductor in the magnetic field may be a driving force for propagation of organic molecules within the

---

[2] http://pds-rings.seti.org

[3] saturn.jpl.nasa.gov/multimedia/images/index.cfm

interstellar space by electromagnetic means, and organic molecules can also be contained within the rings of Saturn, as it is presented in [45-47].

The author would like to express greatest thanks for valuable discussions to A.M. Prokhorov, V.N. Strakhov, V.V. Migulin, Yu.V. Gulyaev, V.I. Pustovoit, A.A. Rukhadze, B.I. Rabinovich, V.G. Kurt, A.Yu. Pospelov, A.N. Malov, O.N. Rzhiga, V.A. Miliaev, E.V. Chensky, V.N. Lugovenko, S.V. Girich, M.V. Belodedov, E.P. Bazhanov, A.V. Zaitsev from the Russian Academy of Sciences; S.V. Vasilyev (SVVTI, CA), V.P. Vasilyev (SOLERC, Ukraine), J.R. Whinnery and T.K. Gustafson (UC Berkeley); J.A. Kong (MIT), D. Osheroff (Stanford), E.A. Marouf (SJSU); J.N. Guzzi, R.B. Hoover, J.F. Spann, R. Sheldon, D. Gallagher, K. Mazuruk and A. Pakhomov (NASA); P. Goldreich (Caltech), L. Spilker (JPL), C.T. Russel and Y. Rahmat-Samii (UCLA), G. Shoemaker (CSUS), L.N. Vanderhoef, P. Rock, R.T. Shelton, R.R. Freeman, A. Albrecht, B. Backer, W. Ko and W. Pickett (UC Davis), R.P. Kudritzki and R.D. Joseph (IFA UH), J.A. Burns (Cornell University, NY), M. Pardavi-Horvath (GTU), P.M. Cincotta (IAFE, Buenos Aires), G. Gerlach (Oriongroup, CA), A. Mendis, L. Peterson, M. Fomenkova (UCSD), N. Castle, R.S. Henderson (San Diego, CA), K. Fischer (Los Gatos, CA), C.B. Vesecky (Palo Alto, CA), A. Pagliere (Sacramento, CA), A.V. Aliaev, V.V. Tsykalo, P.A. Razvin, E.N. Muraviev, O.I. Chernaya (Moscow, Russia) for support. The author is also grateful for fruitful discussions to all participants of the seminars and conferences at the NASA Marshall Space Flight Center and the Huntsville Space Physics Colloquium, the Institute for Astronomy at the University of Hawaii, Astrophysics and the Space Research Center at the University of California in San Diego, CA, the Electrical Engineering and Computer Sciences Department at the University of California in Berkeley, the University of California in Davis, the Institute of Astronomy and Physics at La Plata in Buenos Aires, the Progress In Electromagnetic Research Symposium (PIERS) organized by MIT, the 42nd – 50th SPIE Annual Meetings, the National Bureau of Standard in Washington D.C., the 30th Annual Meeting of the Division of Planetary Sciences of the American Astronomical Society.

REFERENCES

[1] Maxwell, J. C., Brush, S. G., Everitt, C. W. F. and Garber E. (Eds.). (1983). *Maxwell on Saturn's rings*. Cambridge, MA. MIT Press.

[2] Safronov, V. S. (1969). *Evolution of protoplanet cloud and formation of the Earth and planets*. Nauka (Russian), Moscow.

[3] Goldstein, R. M. and Morris G. A. (1973). Radar observations of the rings of Saturn, *Icarus*. V. 20, P. 249-283.

[4] Kaiser, M. L., Desch, V. D. and Lecacheus, A. (1981). Saturnian kilometric radiation: statistical properties and beam geometry, *Nature*. V. 292, P. 731-733.

[5] Brahic, A. (Ed.)., (1984). *Planetary rings. Toulouse: Copadeus.*

[6] Greeberg, R. and Brahic, A. (Eds.). (1984). *Planetary rings*. Tucson: University of Arizona Press.

[7] Bliokh, A. N. and Yaroshenko. V. V. (1991). "Spokes" in the rings of Saturn, *Nature* (Russian). N4, P.19-25.

[8] Spilker, L. J. (Ed.). (1997). *The Cassini-Huygens mission to Saturn and Titan.* Washington, D.C.: JPL, Caltech, NASA SP-533, October.

[9] Rowan, L., Sanchez-Lavega, A., Gombosi, T. I., Hansen, K. S., Porco, C. C., et al, Flasar., F. M., Esposito, L. W. et al., Gurnett D. A. et al., Waite, J. r. et al.,Young, D. T., et al., Dougherty, M. K. et al., Krimigis, S. M. et al and Kempf S. et al. (2005). Cassini at Saturn, *Science*. 25 February. V. 307, №. 5713. P. 1222 – 1276.

[10] Alfven, H. (1981). *Cosmic plasmas.* Dordrecht.

[11] Alfen, H. (1983). Solar system hystory as recorded in the Saturnian rings structure, *Astrophysics Space Science*. V. 97. P. 79-94.

[12] Mendis, D. A., Hill, J. R., Ip, W. H., Goertz, C. K. and Grun, E. (1984). *Electrodynamics processes in the ring system of Saturn*. Saturn, Saturn. T. Gehrels, M. Mathews (Eds.). Tucson: University of Arizona Press. P. 546-589.

[13] Esposito, L. W., Cuzzi, J. N., Holberg, J. B., Marouf, E. A., Tyler, G. L. and Porco, C. C. (1984) *Saturn's rings, structure, dynamics and particle properties*, Saturn. Gehrels T., Matthews M. S. (Eds.). Tucson: University of Arizona Press. P. 463-545.

[14] Gor'kavyi, N. N. and Fridman, A. M. (1994) *Physics of the planetary rings: celestial mechanics of continuous medium*, Nauka (Russian), Moscow. 348 p.

[15] Rabinovich, B. I. (1996). *Magnetohydrodynamic of rotating vortex rings with magnetized plasma*, DAN (Russian). V.351, N3, P.335-338.

[16] Rabinovich, B. I. (1999). *Rotating plasma ring in gravitational and magnetic fields.* Stability problems. DAN (Russian). V.367, N.3, P.345-348.

[17] Ginzburg, V. L. and Andryushin, E. A. (2004). *Superconductivity. World Scientific,* (Please, also see: www.superconductivity.org).

[18] Babushkina, G. V., Kobelev, L. Ya., Yakovlev, E. N. and Babushkin, A. N. (1986). *Superconductivity of ice under high pressure*, Physics of Solid State (Russian). V.28, N12. P.3732-3734.

[19] Maeno, N. (1988). *Science about ice*. Mir (Russian), Moscow. 231 p.

[20] Pospelov, A. Yu, and Tchernyi, V. V. (1995). *Electromagnetic properties material forecast in the planet rings by the methods of functionally physical analysis*, Proc. of International Scientific-Methodical Conference on Innovative Design in Education, Techniques, and Technologies". Volgograd State Technical University, Volgograd, Russia, P. 75-77.

[21] Pospelov, A. Yu., Tchernyi, V. V. and Girich, S. V. (1997). *Planet's rings: super-diamagnetic model and new course of investigations*, Proc. of SPIE 42nd Annual Meeting. San Diego, CA. 27 July-1August . "Small spacecraft, Space Environments and Instrumentation Technologies". V. 3116. № 15. P. 117-128.

[22] Pospelov, A. Yu., Tchernyi V. V., Girich, S. V., Korendovich and V. V. (1997) *Superdiamagnetic model of planetary rings,* Electrodynamics and Techniques of HF and SWF. Moscow. T. 5. № 3. P. 248-262.

[23] Pospelov, A. Yu., Tchernyi, V. V. and Girich, S. V. (1998). *Planetary rings: new mission concept, Kona*, HW: SPIE International Symposiumon Astronomical Telescopes and Instrumentation. 20-28 March. № 132.

[24] Pospelov, A. Yu., Tchernyi, V. V. and Girich, S. V. (1998). *Possible explanation of the planet's rings behavior in the radio and mm-wave range via superdiamagnetic model,* Kona, HW: SPIE International Symposiumon Astronomical Telescopes and Instrumentation. 20-28 March . № 132. № 73.

[25] Pospelov, A. Yu., Tchernyi, V. V. and Girich S.V. (1999). *Possible explanation of the planet's rings behavior in the radio and mm-wave range via superdiamagnetic model //* International Aerospace Abstracts of American Institute of Aeronautics and Astronautics, Inc. January. № 1. P. A99-10781.

[26] Pospelov, A. Yu., Tchernyi, V. V. and Girich S.V. (1998). *Superdiamagnetic model of planetary rings behavior in the millimeter and submullimeter range*, Digest 3465 – 4th International Conference on MM and SMM Waves and Applications. San Diego, CA: Proc. SPIE 43 Annual International Symposium. 20-23 July. San Diego, CA. P. 172-173.

[27] Girich, S. V., Pospelov, A. Yu. and Tchernyi, V. V. (1998). Radar data explanation via superdiamagnetic model of the Saturn's rings, Annual Report of the AAS. 30th Meeting Division of Planetary Science. Madison, WI. 11-16 October. *Bulletin of the American Astronomical Society*. V. 30. № 3. P. 1043.

[28] Pospelov, A. Yu., Tchernyi, V. V. and Girich, S. V. (1999). *Anomalous inversion of polarization of icy satellites and Saturn's rings: superdiamagnetic model,* Proc. 44th SPIE Annual Meeting. Denver, CO. 18-23 July.

[29] (1999). *"Polarization: measurements, analysis and remote sensing II"*. V. 3754. P. 329-333.

[30] Pospelov, A. Yu., Tchernyi, V. V. and Girich, S. V. (2000).*What data could confirm Saturn's rings superconductivity?,* SPIE conference on Astronomical Telescopes and Instrumentation. Munich, Germany. 27-31 March. N. 4015-67.

[31] Pospelov, A. Yu. and Tchernyi, V. V. (2002). *Magnetic levitation of Saturn's rings.,* Progress In Electromagnetic Research Symposium (PIERS). 5-14 July. MIT. Cambridge, MA. C. 135.

[32] Tchernyi, V. V. and Pospelov, A. Yu. (2003). Possible role of space electromagnetism for Saturn's rings existence, *Progress In Electromagnetic research Symposium (PIERS)*. 13-16 October 2003. Honolulu, HW.

[33] Tchernyi, V. V. and Pospelov, A. Yu. (2004). Possible magnetic levitation of Saturn's (planetary) rings. Pisa, Italy, P*rogress in Electromagnetic Research Symposium (PIERS)*. 28-31 March. N. P08.

[34] Pospelov, A. Yu. and Tchernyi, V. V. (2005). Space electromagnetism: modeling of magnetic levitation of superconducting Saturn rings, *Progress In Electromagnetic Research Symposium (PIERS)*. Hangzhou, China. 22-26 August 2005.

[35] Tchernyi, V. V. and Pospelov A.Yu. (2005). Possible electromagnetic nature of the Saturn's rings: superconductivity and magnetic levitation, P*rogress in electromagnetic research (PIER)*. Cambridge, MA: MIT Press. V. 52. P. 277-299.

[36] Tchernyi, V. V. and Pospelov, A. Yu. (2005). *About possible electromagnetic nature of the planetary rings: magnetic levitation of superconducting rings of Saturn*, Fizika volnovykh protzessov i radiotekhnicheskikh sistem (In Russian). T. 8. № 2. P. 4-16.

[37] Tchernyi, V. V. and Chensky, E. V. (2005). Electromagnetic background for possible magnetic levitation of the superconducting rings of Saturn, *Journal of Electromagnetic Waves and Applications*. Cambridge, MA: MIT Press. V. 19. № 15. P. 1997-2006.

[38] Tchernyi, V. V. and Chensky E. V. (2005). *Movements of the protoplanetary sperconding particles in the magnetic field of Saturn lead to the origin of rings,* IEEE Geoscience and remote sensing letters. V. 2. No. 4. P. 445-446. Corrections. IEEE GRSL. 2006. V. 3. No. 2.

[39] Tchernyi, V. V. (Cherny). (2006). *About possible role of electromagnetism and superconductivity for the origin of Saturn's rings,* Prikladnaya fizika (Applied Physics). N. 5. P. 10-16. (In Russian).

[40] Tchernyi, V. V. (Cherny). (2006). *Possible role of superconductivity and electromagnetism for the origin of the rings of Saturn.* Proc. Intern, Conf. "Fundamental principles of engineering sciences". Devoted to 90-years birthday of Nobel prize winner A.M. Prokhorov. Moscow, Oct. 25-27, P.257-259. (In Russian).

[41] Tchernyi. V. V. (2006). Responsibility of electromagnetism for the origin of the rings of Saturn from superconducting particles of the protoplanetary cloud. *//Progress in electromagnetic research symposium (PIERS)*. Tokyo, Japan.

[42] Tchernyi, V. V. and Pospelov, A. Yu. (2007). *A*bout hypothesis of the superconducting origin of the Saturn's rings. Astrophysics and space science. *Springer*. V. 307. No. 4. P. 347-356.

[43] Tchernyi, V. V. (2007). *To the glory of G. Galileo and J.K. Maxwell: electromagnetic modelling of the origin of Saturn's rings from superconducting particles of the protoplanetary cloud*, The 23 rd Annual review of progress in applied computational electromagnetics. Verona, Italy. March 19-23.

[44] Tchernyi, V. V. (2007). *Modeling of electromagnetic origin of the rings of Saturn from superconducting particles of the protoplanetary cloud*, Progress In Electromagnetic Research Symposium (PIERS). Hangzhou, China.

[45] Tchernyi, V. V. and Kapranov, S. V. (2005) Possible role of superconductivity for simplest life propagation within interstellar space by electromagnetic force of magnetic levitation, *Journal of Electromagnetic Waves and Applications.* Cambridge, MA: MIT Press. V. 19. № 15. P. 1997-2006.

[46] Tchernyi, V. V. and Kapranov, S. V. (2005) *Is electromagnetic force a possible means for life transmission in the universe?,* Progress In Electromagnetic Research Symposium (PIERS). Hangzhou, China. 22-26 August.

[47] Tchernyi, V. V. and Kapranov, S. V. (2005) *Contribution of superconductivity to possible interstellar propagation of organic molecules by electromagnetic way,* Proc. 50th SPIE Annual Meeting. July 31-Aug. 4. San Diego, CA. Hoover R.B. et al. (Eds.). Astrobiology and planetary missions. SPIE. V. 5906.

In: New Developments in Magnetohydrodynamics …
Editors: Stephen Talton and Patrick F. Theba
ISBN: 978-1-61942-078-6
2012 Nova Science Publishers, Inc.

Chapter 3

HIGH SPEED FLOWFIELD ANALYSIS FOR SATELLITE LAUNCH VEHICLE AND REENTRY CAPSULE

R. C. Mehta [*]
School of Mechanical and Aerospace Engineering
Nanyang Technological University, Singapore

ABSTRACT

The chapter presents numerical and experimental analyses of satellite launch vehicle and reentry capsule at high-speeds. Numerical simulations of free jets emanating from nozzle, impinging of supersonic jets on jet deflector, flowfield over bulbous payload shroud in the transonic and supersonic Mach number range, drag reducing spike attached to blunt body, prediction of angle of attack from the flight data of launch vehicle and depressurization of heat shield during ascent phase of flight. Influences geometrical parameters of various types of reentry capsules on aerodynamic drag are studied at various trajectory points. Results of computational fluid dynamics are compared with the experimental and flight data and exhibit fairly good agreement between them. The reliable estimation of aerodynamic problem associated with launch vehicle can be made in conjunction with computed flow visualizations.

INTRODUCTION

The satellite launch vehicle needs flowfield characteristics during lift-off of the launch vehicle, in the transonic range and at the maximum dynamic pressure condition, depressurization of heat shield compartment during the ascent phase of the flight. Forward facing spike is attached to blunt body to decrease the aerodynamic drag. In the case of reentry capsule the blunt cone radius is increased to decelerate the reentry module. The post flight analysis of launch vehicle requires information of angle of attack for evaluating the aerodynamic and structural performance of the vehicle. The chapter presents numerical simulations with the corresponding experimental data. The chapter deals the following

[*] Email address: RC Mehta@ntu.edu.sg

aerodynamic problems related to the satellite launch vehicle and the reentry capsules at high-speeds.

Diffusers, free jets, and impinging jets problems find many applications in aerospace industries. The flowfield produced by expanding supersonic emanating from a rocket nozzle produces very complicated flowfield characteristics. The structures of the flowfield consist of jet shocks, reflecting shocks, Mach discs, jet boundaries, slip lines, and mixing regions etc. Mechanisms for the occurrence of Mach disks in the jet are not satisfactorily understood.

A bulbous payload shroud is generally selected to accommodate increased volume of the satellite in a launch vehicle. One of the serious problems of the bulbous heat shield is the wall pressure fluctuations caused by shock wave-turbulent boundary layer interaction. The estimation of flowfield characteristics around such configurations is of practical importance, as well as research interest. The occurrence of buffeting during atmospheric ascent depends primarily on the shape of the heat shield. To eliminate the buffeting problem, the NASA guidelines recommended certain values of cone angle, cone and cylinder lengths, and boat-tail angle of the heat shield.

A high-speed flow past a blunt body generates a bow shock wave which causes a rather high surface pressure and as a result the development of a high aerodynamic drag. The pressure on the blunt body surface can be substantially reduced if instead of a normal shock in the stagnation region, an oblique shock is generated by aerospike. The separated flow in the aerospike region produces self-oscillating flow which has to be investigated numerically and experimentally.

During the atmospheric flight of satellite launch vehicle and sounding rocket, the compartment of the vehicle needs depressurization to prevent abnormal pressure built-up, which can be detrimental to the vehicle. The differential pressure depends mainly on the location of vent ports, the effective volume of air to be evacuated, the trajectory of the launch vehicle. The NASA monograph describes the design criteria of the compartment venting for space vehicle during ascent and reentry phase.

Highly blunt configuration of reentry capsules are generally preferred to decelerate spacecraft for safe returning on earth after performing the experiments. The primary design requirement of reentry capsules are large nose-radius for high aerodynamic drag and short body length for reducing the total structural weight, resulting in a small ballistic coefficient. The flowfield over the reentry module becomes complicated due to presence of corner at the shoulder and base-shell of the capsule.

The above mentioned aerodynamic problems related to satellite launch vehicle and reentry capsules are analyzed using numerical approach. The experimental data are used to validate the computed results. Many interesting flowfield features are revealed using the numerical analysis and flowfield visualization of Mach and density contour plots.

1. Flow Field of Supersonic Free Jets and Jet Deflector

The phenomenon of supersonic jets and their interaction with deflector surfaces is found many engineering applications such as impingement of exhaust from a rocket nozzle of launch vehicles during the liftoff phase, during stage separation of multistage rockets, and VTOL/STOL operation of aircraft etc. An interaction between a supersonic jet and a flat plate at atmospheric conditions has been studied [1] as the fundamental problem to the impingement of a rocket exhaust plume on the deflector or plate that of a control jet on the rocket body, etc. The impingement jet is characterized by many discontinuities such as barrel shock, exhaust gas jet boundary, Mach disk, reflected shock, plate shock. The flowfield analysis of free jets impinging over jet deflector is required in order to know the thermal, pressure and acoustics loads. The flow structure of an underexpanded supersonic jet impinging on an inclined plate has been numerically investigated by Kim and Chang [2] by solving the compressible Euler equations. It is found by them that the inviscid numerical simulation can offer fairly good prediction of the impinging jet for moderate plate inclination. The maximum pressure was found to be larger on the inclined plate than on the normal plate. Flow fields of the exit jet Mach number of 2.2 impinging on an inclined plate at various plate angles, nozzle-plate distance and pressure ratios are experimentally investigated using pressure-sensitive paints and schlieren flow visualization [3]. The flowfield patterns can be predicted without experiments if the shock cell length in the free jet for various pressure ratios is known in advance. Numerical study of under expanded jet impinging against the solid wall has been presented by Wu et al. [4]. The acoustic efficiency of the deflected jet exhaust on a curved surface is less than the jet impinging on a flat plate [5].

Flow field features of supersonic free-jets as well as impinging jets over the axisymmetric and double wedge deflector are numerically simulated. The differences of the flowfield behavior over axisymmetric and double wedge jet deflectors are analyzed in Refs. [6] and [7]. The numerical studies [8] of the flowfield structure in diffusers, free jets and impinging jets have been investigated for a range of jet-to-stream total pressure ratios and for different exit Mach numbers. The numerical simulation has shown an oscillating phenomena [8] in the case of free jets operating at sonic exit Mach number and in another case of impinging of the under expanded jets on a flat plate. Effects of Mach number and exit to ambient pressure ratio on shock cell lengths are investigated with the aid of schlieren pictures and density contour plots by Mehta and Prasad [9]. Figure 1 illustrates, the under expanded flow produced by a supersonic nozzle. L_1, L_2 and L_3 are the first, second and third shock cell length, respectively, as delineated in the schematic sketch of a supersonic free jet. The first shock cell length of slightly incorrectly expanded axisymmetric jet was theoretically estimated by Prandtl to vary as $L_1/D_e = k\beta$, where $k = 1.306$ and D_e is nozzle exit diameter as mentioned in Ref. [10]. The parameter β is defined in Eq. (1). Love et al. [11] presented semi empirical expressions for calculation of first shock cell length. Abdel-Fattah [12] has measured shock cell lengths for supersonic jets issuing from a conical convergent-divergent nozzle in conjunction with schlieren pictures. They observed from schlieren photographs that the first and second shock cell lengths are having a linear dependence on β^2 rather than β.

Table 1. Nozzle Dimensions

M_e	D_e, mm	D^*, mm
2.2	30.0	21.18
2.6	14.6	9.20
3.1	23.0	10.65

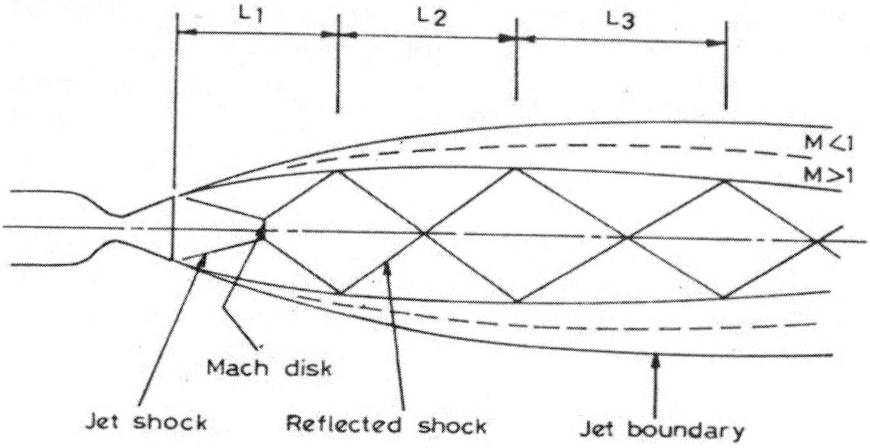

Figure 1. Schematic sketch of a supersonic free jet.

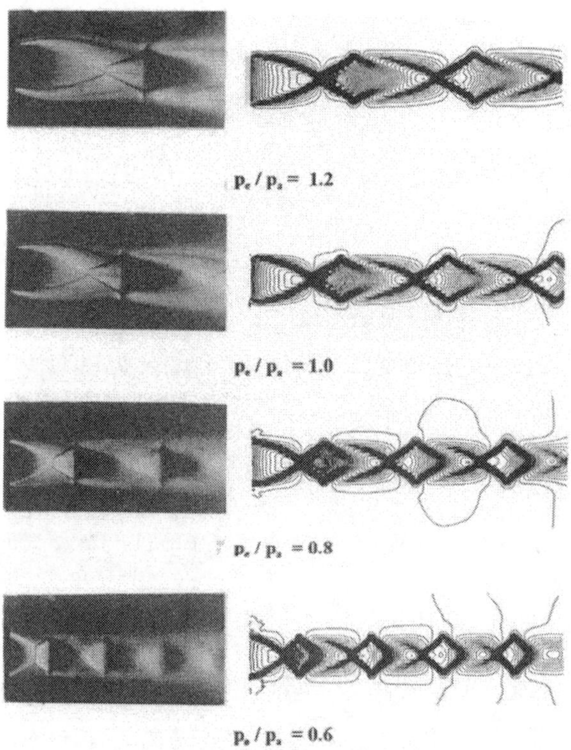

Figure 2. Schlieren photographs and pressure contours at different p_e/p_a for $M_e = 2.2$.

Numerical and experimental studies [13] are carried out for supersonic free jets operation various nozzles exit Mach and exit to ambient pressure ratio. The convergent-divergent nozzles were designed for producing supersonic free jets. D^* is the throat diameter of the nozzle. The exit diameter is designated as D_e and has a semi-divergent angle of 15°. The particulars of the nozzles are as given in Table 1. The D^* and D_e represent the diameter of the throat and nozzle exit, respectively. A static pressure probe of 40^0 sharp cone in the front end followed by pressure ports at a distance of 4.89 mm from the nose tip is used to measure the static pressure in the free jets. In the static probe, four holes are situated at 90^0 apart and are pneumatically connected. The pressure probe is calibrated and is found to measure static pressure within ±1% of nominal value. The hysteresis, reproducibility, and uncertainty of the measurements are carried out. The geometrical detail of the static probe is given in Ref. [13]. A two-stage Runge-Kutta time-stepping method in conjunction with finite volume discretization scheme is applied to solve inviscid, axially symmetric flow. The schlieren pictures and density contour plots exhibits all the essential flowfield features of the supersonic free jets, such as jet boundary, jet shock, Mach disk and reflected shock. The schlieren pictures of free jets and corresponding isobar contours at jet exit Mach number of 2.2 at various jet exits to ambient pressure ratio (p_e/p_a) are depicted in Figure 2. The numerical and the schlieren pictures reveals that for a constant exit Mach number, the location of Mach disk moves closer to the nozzle exit plane and the length of the shock-cell length decreases with decreases in the exit to ambient pressure ratios. Moreover, the location of the Mach disk moves away from the nozzle exit plane as the exit Mach number increase while taking same value of (p_e/p_a).

The shock cell lengths were obtained from the numerical and schlieren pictures for jet exit Mach number of 2.20, 2.61 and 3.10 at various jets exit to ambient pressure ratio (p_e/p_a). The L_1/d^* and the L_2/d^* versus β^2 are obtained by solving compressible, axisymmetric Euler equations and the schlieren pictures [9], where β^2 is defined as

$$\beta^2 = \left(\frac{2}{\gamma-1}\right)\left(\frac{P_0}{p_e}\right)^{\frac{\gamma-1}{\gamma}} - \left(\frac{\gamma+1}{\gamma-1}\right) \tag{1}$$

where P_0 is the jet stagnation pressure, p_e is the nozzle exit pressure, γ is ratio of the specific heats. The first and second shock cell lengths are measured employing schlieren pictures and density contour plots. The following linear relationship [9], [13] are obtained by applying least square fit for each nozzle

$$L_1/D^* = 0.95\beta^2 - 0.80\text{, for } M_e = 2.20,\ 2.2 < \beta^2 < 3.6$$

$$L_1/D^* = 1.06\beta^2 - 1.08\text{, for } M_e = 2.61,\ 3.8 < \beta^2 < 5.8$$

$$L_1/D^* = 1.25\beta^2 - 3.35\text{, for } M_e = 3.10,\ 5.7 < \beta^2 < 8.7 \tag{2}$$

The validation of the constants is checked by directly comparing with the results of Abdel-Fattah [12]. The commutated values of second shock cell length from the numerical results show following linear relationship

$$L_2/D^* = 0.72\beta^2\text{, for } 2.2 < Me < 3.10 \tag{3}$$

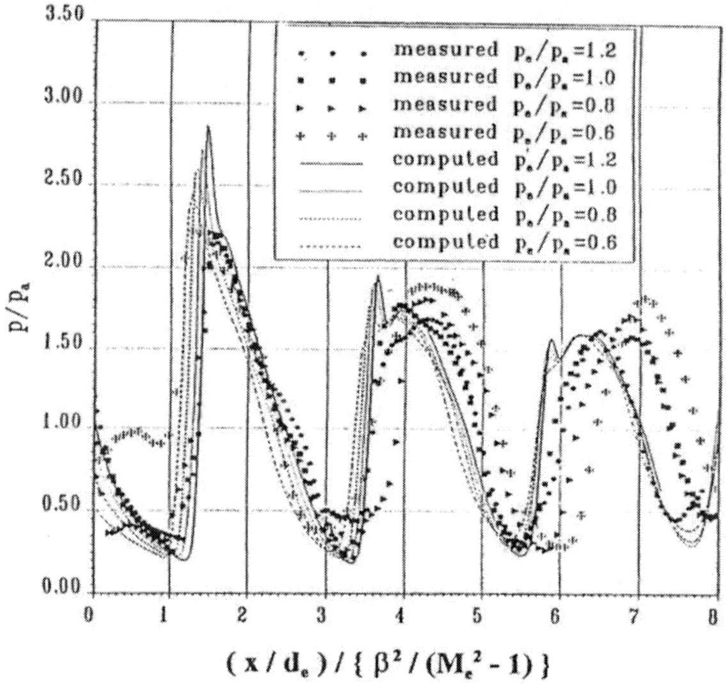

Figure 3. Pressure variation along the jet axis for $M_e = 2.2$ at different values of p_e/p_a.

The second shock cell length became independent of exit Mach M_e. The difference between experimental with computed results may be attributed to the faster jet spreading rate and viscous effects after first shock cell and hence the lower length of second shock cell length. Equations (2) and (3) show a linear relation for first and second shock cell length as function of β^2, which has been obtained for nozzles of different exit Mach number.

Figure 3 shows the variation of static pressure along the jet axis for $M_e = 2.2$ at different values of p_e/p_a. The pitot p_{t2} and the static pressure distribution p_s ore measured and the distribution of the local Mach number in the free jets are computed by solving following isentropic equations [14]

$$\frac{p_{t2}}{p_s} = \left[1 + \frac{(\gamma - 1)}{2} M^2\right]^{\frac{\gamma}{\gamma - 1}} \quad M < 1 \tag{4}$$

and

$$\frac{p_{t2}}{p_s} = \left[1 + \frac{(\gamma - 1)}{2} M^2\right]^{\frac{\gamma}{\gamma - 1}} \left[\frac{(\gamma - 1)}{2\gamma M^2 - (\gamma - 1)}\right]^{\frac{1}{\gamma - 1}} \quad M \geq 1 \tag{5}$$

The measured pitot and static pressure distributions have been used to obtain the exit Mach of M_e = 2.2, 2.6, and 3.1 at various values of jet exits to ambient pressure ratio. An iterative scheme is used solve Eqs. (4) and (5) to get the value of Mach number variation along the jet axis. A parameter $[\beta^2/(M_e^2 - 1)]$ has been obtained to characterize the distribution of pressure and also Mach number along the jet axis which makes them invariant of the exit to ambient pressure ratio. The variation of the Mach number along the jet axis and versus non-dimensional parameter $[\beta^2/(M_e^2 - 1)]$ also makes Mach number independent of exit to ambient pressure ratio (p_e/p_a) as shown in Refs. [9] and [15]. The pitot pressure remains constant in the flow immediately downstream of the Mach disk, which is attributed to subsonic and inviscid flow. The pitot pressure then rises owing to the inward diffusion of high pressure from the outer part of the jet.

The jet mixing noise level and spectral character is a function of the velocity differential between that of the jet and its environment, the jet density relative to the density of the surrounding air and the jet dimensions. The sound pressure level for fully expanded jet density is function of cross-sectional area of jet exhaust nozzle, ambient-total pressure, static pressure under sea level conditions, and radial distance from nozzle exit to observer. An incorrectly expanded supersonic jet contains shock waves which interact with jet turbulence to produce a source of noise in addition to due to the turbulent mixing process. When a convergent nozzle is operated at a supercritical pressure ratio, a regular shock structure exists in the resulting jet flow. This structure gives rise to shock noise in addition to the basic jet mixing noise. The shock related noise can be divided into two distinct types, each having its own characteristics properties. The first component is discrete in nature, usually with several harmonics, and it is often referred to as the "screech" component. The second component is broadband in nature with well-defined maximum at a specific frequency. It is this broadband shock associated noise that is of real significance in the farfield noise. The jet mixing noise and the shock noise from a supersonic jet cannot be separated explicitly. Powell [16] developed a model based on a feedback mechanism between the shock structure and the nozzle lip for the screech component of shock noise.

The geometrical detail of the axisymmetric model depicted in Figure 4 is employed to study the numerical and the experimental simulation of the supersonic jet impinging on a typical jet deflector. Measured and calculated pressure distribution [9] and Mach number [14] along the jet axis for various operating conditions indicate that a nondimensional parameter could be obtained to make axial pressure distribution independent of exit to ambient pressure ratio.

The axisymmetric jet deflector model consists of a cone-apex angle of 70o and a tip bluntness radius of R of 0.13 De is provided at a radial location r = 0.7 De, where r is the distance measured from the deflector axis. The model has a base diameter of 0.8 De. The deflector was carefully aligned with its base perpendicular to the jet axis and the apex was carefully located. The process of centralizing the model was accomplished by equalizing the reading from pressure holes at corresponding position on opposite faces of the deflector. The static pressure distribution in the axial direction of free jets was measured using a miniature static probe [13]. The measurement of surface pressures on the jet deflector is done by pressure transducers. Figure 5 shows comparison between density contour plot and schlieren picture over the axisymmetric jet deflector. The density contours plots shows all the essential flow field features like jet shock, jet boundary, Mach disk, reflected shock and cone shock. The comparison of the schlieren pictures of the free jets and impinging jets reveals that the impinging flow field has similar characteristics of the free jet up to the cone shock of the deflector's apex. However, the vector plots of the impinging free jets differ considerably from the impinging vector plots [6]. The difference in the velocity field may be attributed to the formation of the vortex. Centre line pressure variation in the presence of the deflector is shown in Figure 6 at different operating condition of the nozzle. The static pressure distribution on the centre line of the deflector with respect to nozzle is nondimensionalized by the settling chamber pressure P0, and x is nondimensionalized by the nozzle exit diameter De. A sharp rise of the pressure on the deflector (stagnation region) is reported in Ref. [17].

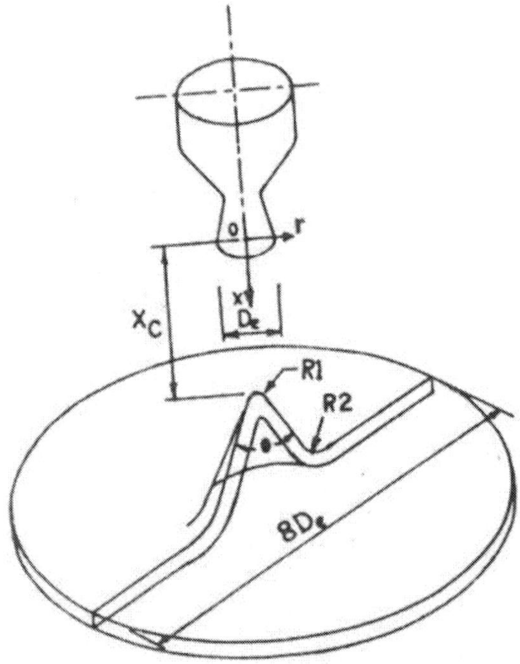

Figure 4. Coordinate system and axisymmetric model configuration.

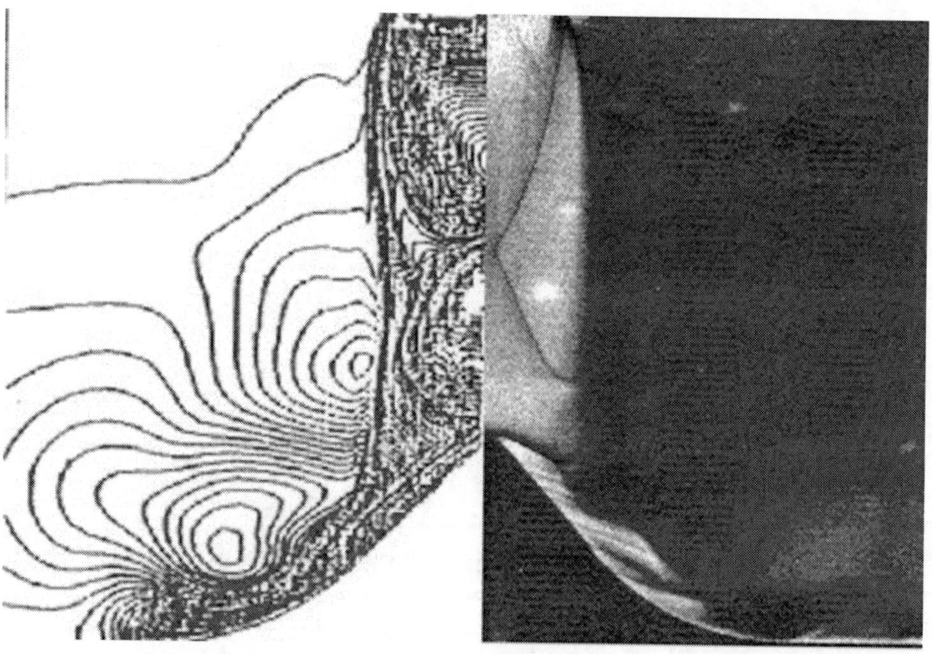

Figure 5. Comparison between density contour plots and schlieren picture over jet deflector ($M_e = 2.2$, $X_c/D_e = 3.0$).

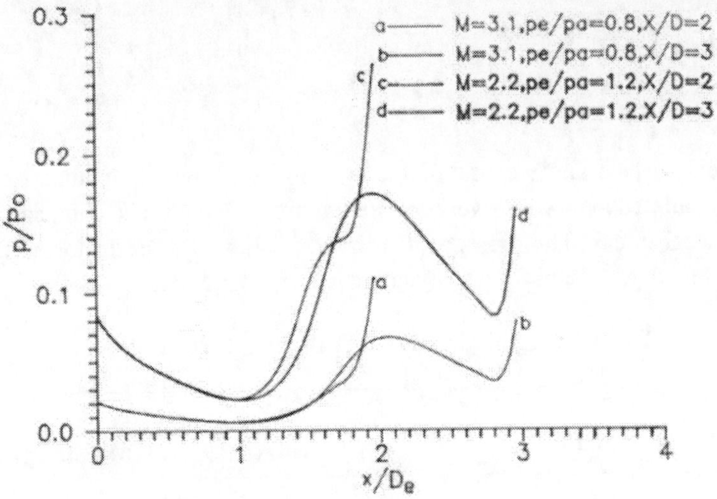

Figure 6. Centre line pressure distribution in the presence of axisymmetric jet deflector.

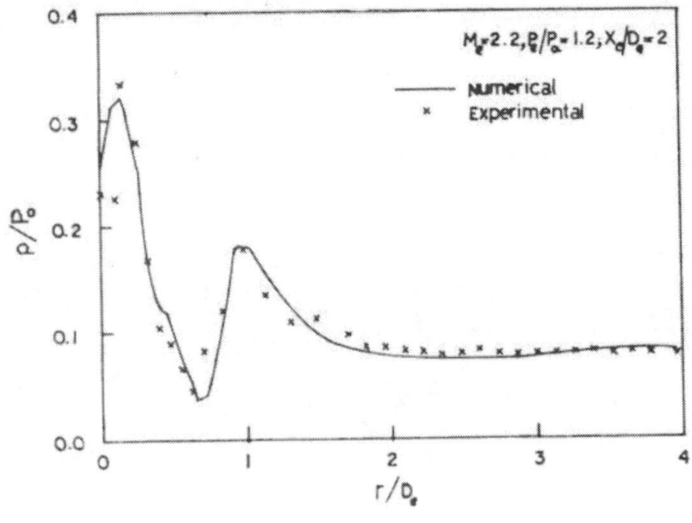

Figure 7. Surface pressure distribution over axisymmetric jet deflector (Me = 2.2, pe/pa 1.2, Xc/De = 2.0).

The centre line variation inside the jets differs in the presence of the jet deflector with the distance, X_c and the nozzle exit Mach number, M_e. Figure 7 shows pressure distribution on the jet deflector for $M_e = 2.2$, $p_e/p_a = 1.2$ and $X_c/D_e = 2.0$. It can be seen from the pressure distribution a sharp rise of the pressure on the apex of the deflector and second peak in the pressure is found on the curvature region. The inviscid simulation of the fluid dynamics equations compares fairly well with the experimental data.

A quantity, which is of practical importance is the overall impinging jet loads on the jet deflector surface. The axial load is required for the structural design of the jet deflector. Loads have been obtained by integrating the measured pressure distribution. The axial load distribution C_a over the axisymmetric jet deflector is computed using following relation:

$$C_a = \frac{2\pi\int_0^{4D_e}(p - p_a)r\sin\psi ds}{P_o A_e} \tag{6}$$

where ψ is local surface angle of the deflector. Table 2 shows the comparison between the measure and calculated axial load over the axisymetric deflector. The comparison shows good agreement between them. The pressure distributions above the deflector surface as well as integrated axial load are useful for aerodynamic and structural design.

Table 2. Axial load Over Axisymmetric Deflector

M_e	p_e/p_a	X_c/D_e	C_a	
			Calculated	Measured
2.2	1.2	2	0.70	0.74
2.2	1.2	3	0.65	0.70
3.1	0.8	2	0.75	0.82
3.1	0.8	3	0.63	0.73

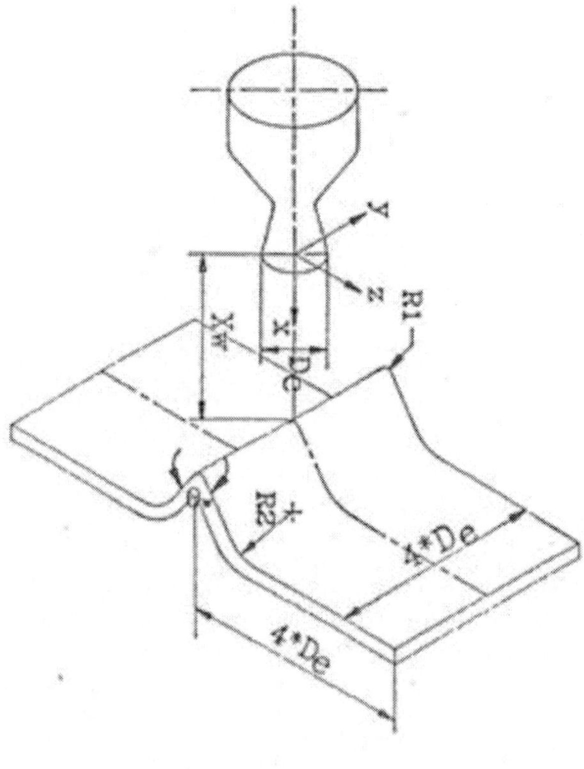

Figure 8. Geometrical detail of the wedge deflector model.

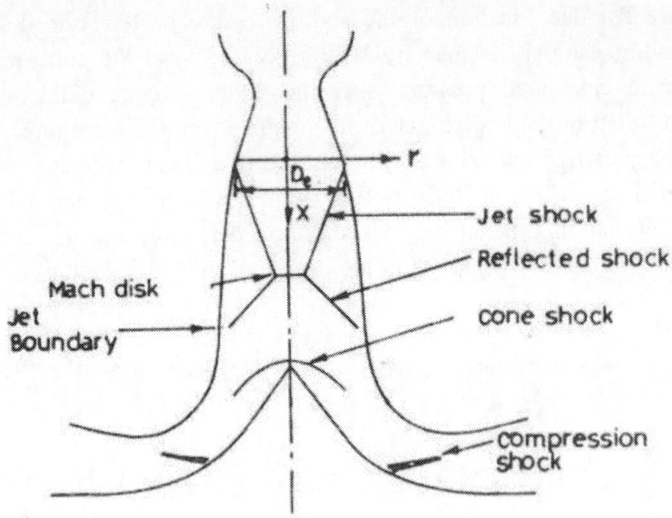

Figure 9. Schematic of the impingement flow field over jet deflector.

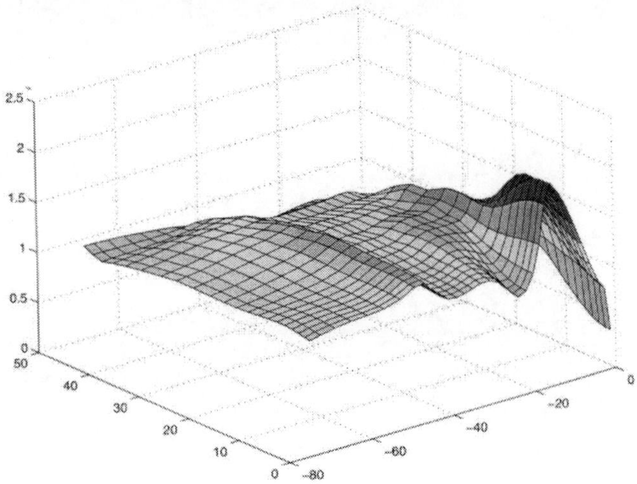

Figure 10. Surface Mach distribution over wedge deflector (Me = 2.2, $X_c/D_e = 2.0$).

Figure 8 gives geometrical details of the double wedge deflector model. The wedge deflector has the same apex angle, radius of curvature and base length as used in the axisymmetric jet deflector. The impingement jet is on the wedge deflector characterized by many discontinuities such as barrel shock wave, exhaust gas jet boundary, Mach disk, reflected shock, plate shock as delineated in schematic sketch of the flowfield based on the experimental data in Figure 9. The flowfield over the double wedge deflector is obtained by solving three-dimensional compressible Euler equations. Carpet plot of surface Mach distribution over wedge deflector for exit Mach number $M_e = 2.2$, $X_c/D_e = 2.0$ is shown in Figure 10. The Mach contour plot is shown in Figure 11 at the centre plane of the impingement jet. It can be seen from the Mach contour plots that the essential flow features of the impinging supersonic jets on the double wedge deflector are captured very well as shown in Figure 9. The static pressure p on the deflector surface is normalized by the settling chamber pressure P_0. The location of the static pressure port on the deflector surface is indicated as nondimensional distance z_e/D_e. Figure 12 shows the comparison between the measured [18] and calculated pressure over the double wedge deflector. The measured surface pressure distribution were used to obtain overall impinging load coefficient. Integration has been carried out using the trapezoidal rule to calculate the load coefficient and results shows good agreement in Ref. [7]. The numerical and experimental investigations of supersonic free jets, impinging on the axisymmetric deflector, and the double jet deflector show good agreement can be employed in the design of the jet deflector during the liftoff the launch vehicle.

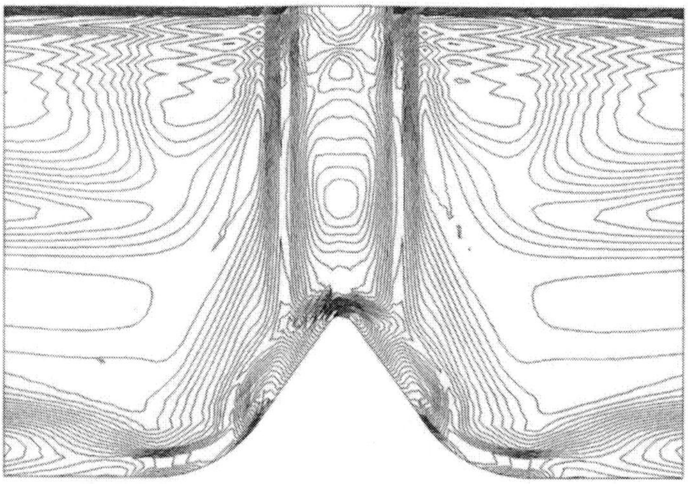

Figure 11. Mach contour plots over wedge deflector ($M_e = 2.2$, $X_c/D_e = 2.0$).

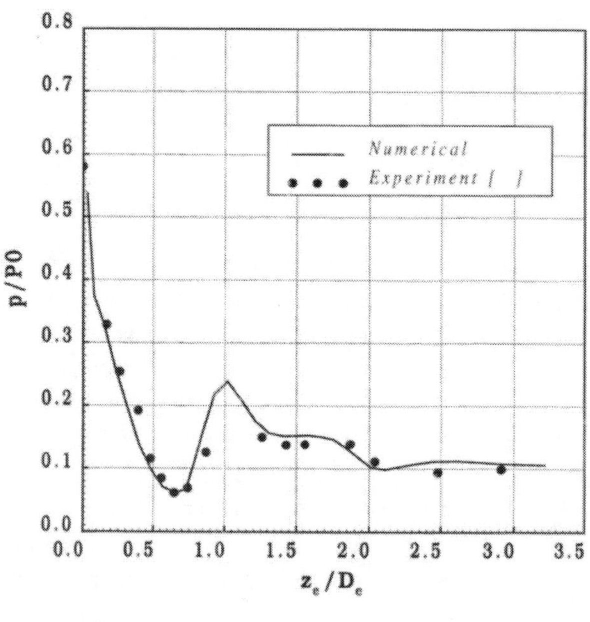

Figure 12. Centre line static pressure distribution over the deflector ($M_e = 2.2$, $X_c/D_e = 2.0$).

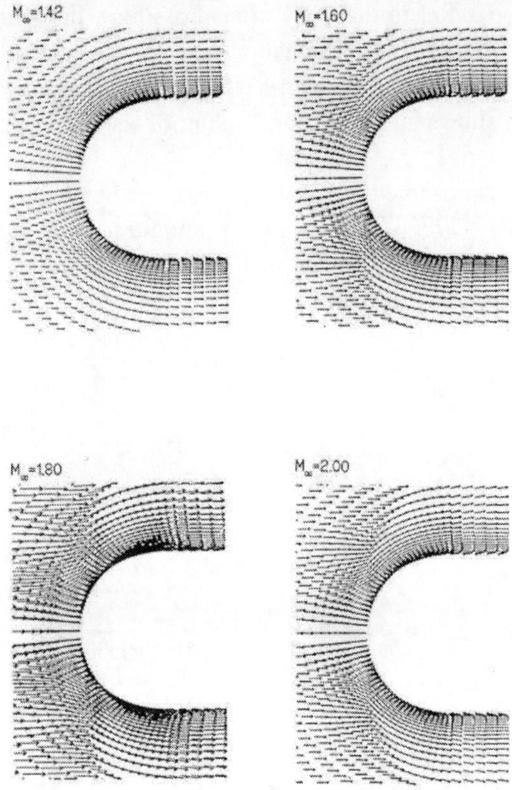

Figure 13. Computed velocity field and shock position over a hemisphere-cylinder.

2. Computation of Flow Field over Payload Faring

The design process for aerospace application has been improved significantly by the use of computational fluid dynamics. Numerical solutions of the Navier-Stokes equations have been used to obtain a better understanding of the quantitative and qualitative physical phenomena in viscous flows. The application of a hemisphere-cylinder is one of the basic nose configurations for blunt nose body of revolution. A detailed analysis of the phenomena of viscous flow will reveal the physical flowfield features of transonic and supersonic flow past nose-body in general.

Hsieh [19] and [20] has conducted wind-tunnel tests of a hemisphere-cylinder model of $2.54 x 10^{-3}$ m diameter at zero angle of incidence and Mach number range of 0.7 – 1.3. The time-dependent axisymmetric Reynolds-averaged turbulent compressible Navier-Stokes equation is solved using finite volume scheme in conjunction with the multi-stage Runge-Kutta time stepping method. The closure of the system of equations is achieved by introducing the turbulence model of Baldwin-Lomax [21]. The bow shock stand off distance is calculated and compared with the available data in Ref. [22]. Figure 13 displays an enlarged view of the computed velocity field at supersonic Mach number range of 1.4 – 2.0. It can be easily seen that the bow shock wave moves close to the body with increasing Mach number which is also coinciding with the shock position. It appears from the velocity field that the flow becomes parallel to the body after crossing a line perpendicular to the model axis. This line is in the vicinity of the stagnation point of the hemisphere. A schematic sketch of the flow pattern is delineated in Figure 14 which is constructed based on the observation. The numerical analysis shows an interesting finding of the flow past hemispherical-cylinder body [22].

The separated flow is observed at the shoulder of the hemisphere-cylinder for freestream Mach number of 0.80 – 0.90. Unsteady flow simulation over the hemisphere-cylinder is carried out at transonic Mach number range. The maximum value of the root-mean-square wall pressure fluctuations [23] is observed at $M_\infty = 0.814$.

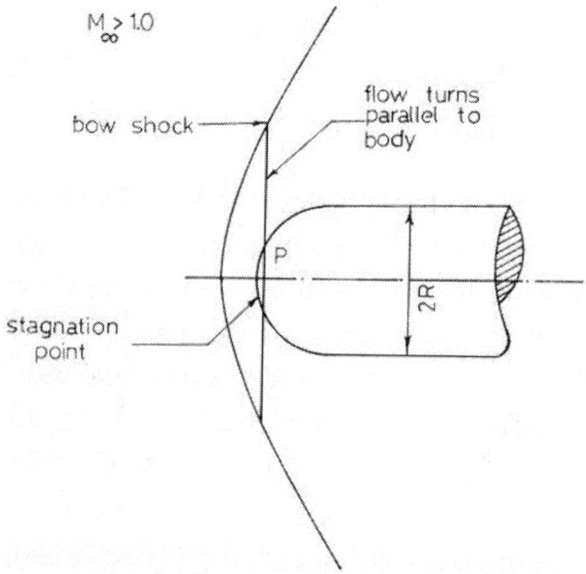

Figure 14. Schematic sketch of the flow over a hemisphere-cylinder.

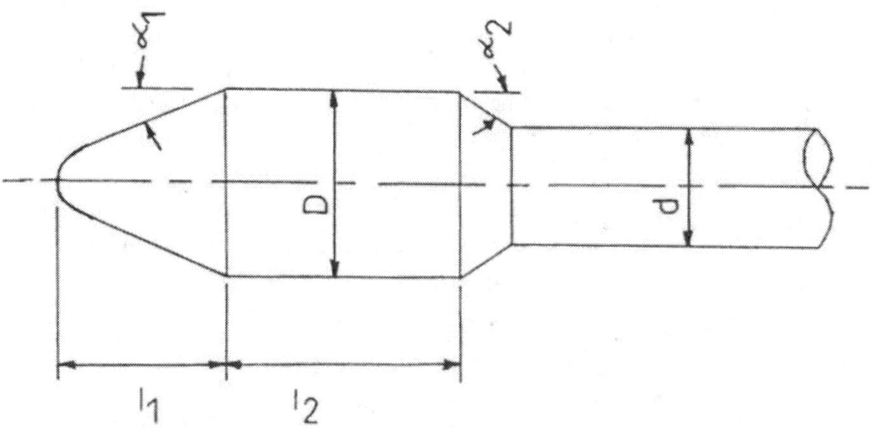

Figure 15. Geometrical parameters of heat shield.

Table 3. Geometrical parameters of heat Shield

Model	α_1, deg	L_1/D	L_2/D	d/D	α_2, deg
M1	20	0.965	1.400	0.875	15
M2	15	1.023	1.117	0.823	30
NASA	≤ 15	≥ 0.80	> 1.50	0.9 - 1.11	Not critical

The bulbous payload shroud is generally selected to accommodate increased volume of the satellite in a launch vehicle. One of the serious problems in high-speed flight of the bulbous payload shroud is wall pressure fluctuations caused by shock-wave turbulent boundary layer interaction. The estimation of flowfield behavior around such configurations is very practical, as well as research interest. A shock wave of sufficient strength interacting with boundary layer can cause separation and process may become unstable. The strength of the terminal shock and the mechanism of their interaction are linked to a specific configuration of heat shield of a satellite launch vehicle. Interaction unsteadiness may produce large amplitude fluctuations loads. The frequency band of the acoustic loads is typically in the range of several hundred Hz to several kHz. Table 3 and Figure 15 depict the dimensions of the heat shield considering for the numerical simulation with the recommended NASA guidelines [24] for the buffet-free configuration. A numerical simulation of the unsteady, compressible, turbulent, axisymmetric Reynolds-averaged Navier-Stokes (RANS) equations is attempted in order to understand the basic fluid dynamics. The governing equations can be written in the following strong conservation form

$$\frac{\partial \mathbf{W}}{\partial t}+\frac{\partial \mathbf{F}}{\partial x}+\frac{\partial \mathbf{G}}{\partial r}+\mathbf{H}=\left[\frac{\partial \mathbf{R}}{\partial x}+\frac{\partial \mathbf{S}}{\partial r}\right] \tag{7}$$

where

$$\mathbf{W}=r\begin{bmatrix}\rho\\ \rho u\\ \rho v\\ \rho e\end{bmatrix},\quad \mathbf{F}=r\begin{bmatrix}\rho u\\ \rho u^2+p\\ \rho uv\\ (\rho e+p)u\end{bmatrix},\quad \mathbf{G}=r\begin{bmatrix}\rho v\\ \rho uv\\ \rho v^2+p\\ (\rho e+p)v\end{bmatrix} \tag{8}$$

are the W is conservative state and F and G are inviscid flux vectors. x and r are axial and radial coordinate system and t is time. The viscous flux vectors are

$$\mathbf{R}=r\begin{bmatrix}0\\ \sigma_{xx}\\ \tau_{xr}\\ u\sigma_{xx}+v\tau_{xr}+q_x\end{bmatrix},\quad \mathbf{S}=r\begin{bmatrix}0\\ \tau_{xx}\\ \sigma_{xr}\\ u\tau_{xr}+v\sigma_{rr}+q_r\end{bmatrix},\quad \mathbf{H}=\begin{bmatrix}0\\ 0\\ \sigma_{+}\\ 0\end{bmatrix} \tag{9}$$

where R and S are viscous flux vector and S is source vector. σ and τ are components of the stress vector, while q is components of the heat flux vector. Reynolds stresses and turbulent heat fluxes in the mean flow equations are modeled by introducing an isotropic eddy viscosity μ_t. Thus, the viscous terms in the equations became

$$\sigma_{xx} = -\frac{2}{3}(\mu + \mu_t)\nabla \cdot U + 2(\mu + \mu_t)\frac{\partial u}{\partial x}$$

$$\sigma_{rr} = -\frac{2}{3}(\mu + \mu_t)\nabla \cdot U + 2(\mu + \mu_t)\frac{\partial v}{\partial r}$$

$$\tau_{xr} = (\mu + \mu_t)\left(\frac{\partial u}{\partial r} + \frac{\partial v}{\partial x}\right)$$

$$\sigma_{+} = -p - \frac{2}{3}(\mu + \mu_t)\nabla \cdot U + 2(\mu + \mu_t)\frac{v}{r} \tag{10}$$

$$\nabla \cdot U = \frac{\partial u}{\partial x} + \frac{\partial v}{\partial r} + \frac{v}{r}$$

$$q_x = -cp\left(\frac{\mu}{\Pr} + \frac{\mu_t}{\Pr_t}\right)\frac{\partial T}{\partial x}$$

$$q_r = -cp\left(\frac{\mu}{\Pr} + \frac{\mu_t}{\Pr_t}\right)\frac{\partial T}{\partial r}$$

where Pr and Pr_t, the molecular and turbulent Prandtl numbers are assumed to take a constant value of 0.72 and 0.90, respectively. cp is specific heat at constant pressure. The coefficient of molecular viscosity is calculated according to Sutherland's law

$$\mu = 1.458x10^{-6}\frac{T^{1.5}}{T + 110.4} \tag{11}$$

The temperature is related to pressure and density by perfect gas equation of state as

$$p = (\gamma - 1)\left[e - \frac{1}{2}\rho(u^2 + v^2)\right] \tag{12}$$

The ratio of the specific heat γ was assumed constant and equal to 1.4. The closure of the system of equations is achieved by introducing the Baldwin-Lomax turbulence model [21]. This model has been extensively used previously and is reputed to yield acceptable engineering solutions.

In the inner region the eddy viscosity is given by

$$(\mu_t)_i = (0.4D_1L)\rho|\omega| \tag{13}$$

where ω is the vorticity function, L is the normal distance to the nozzle wall and D_1 is Van Driest's damping factor

$$D_1 = 1 - \exp\left[-\left(\frac{\rho_w |\omega|_w}{\mu_w} \right)^{0.5} \frac{L}{26} \right] \tag{14}$$

In the outer region,

$$(\mu_t)_o = 0.0168(1.6) F_w F_{KIF} \tag{15}$$

The coefficient Fw is calculated as the minimum of the following two values

$L_{max} F_{max}$,

$0.25 L_{max} \max\left[(u^2 + v^2) \right]^{0.5} / F_{max}$

The scale length L_{max} is the maximum value of L when the function F ($= L D_1 |\omega|$) attains its maximum F_{max}. The Klebanoff intermittency correction factor is given by

$$F_{KIF} = \left[1 + 5.5 \left(0.3 \frac{L}{L_{max}} \right)^6 \right]^{-1} \tag{16}$$

The effective viscosity is then given by

$$\mu_t = \min(\mu_i, \mu_o) \tag{17}$$

The numerical algorithm employs a finite volume discretization technique. The integral conservation equations are applied to each computational cell. Temporal integration is performed using multistage time-stepping method. The dimensional of the heat shield M1 and M2 are given in Table 3 and Figure 15. The maximum diameter of the model is D. The boat-tail angle of heat shield M1 and M2 are 15^0 and 30^0, respectively. For the blunt nosed cone, the semi-cone angle is $20^0$and 15^0 for the models M1 and M2, respectively. Further details of the numerical procedure, initial and boundary conditions, and computational grid arrangement are given in Ref. 25. Figure 16 shows all of the essential flow field features of the transonic flow, such as supersonic pocket, terminal shock, and expansion and compression regions are very well captured and compare well with the schlieren photographs. The density contour plots reveal that the supersonic region increases with increasing freestream Mach number, and as a result, the terminal shock moves downstream with increasing freestream Mach number. It is worth to mention here that the terminal shock is just reached at the end of the payload shroud in the case of M1 heat shield, whereas the terminal shock crosses the fore body cylinder in the M2 configuration at the same freestream Mach number. It can be noticed that the M1 model is able to hold the terminal shock, which is attributed to the greater length of the payload shroud. The flow field characteristics are shown in detailed in Refs. [25 – 28].

Comparison between the shock locations of the numerical and experimental results is shown in Figure 17 for M1 heat shield. It can be seen that the terminal shock crosses the fore body cylinder at $M_\infty = 0.97$. The location of the terminal shock is found nonlinear function of Mach number. The location of the shock is confirmed during the flight of the launch vehicle as depicted in Figure 18. It is interesting to note that the numerical analysis compares well with the wind-tunnel data as well as the flight measurement.

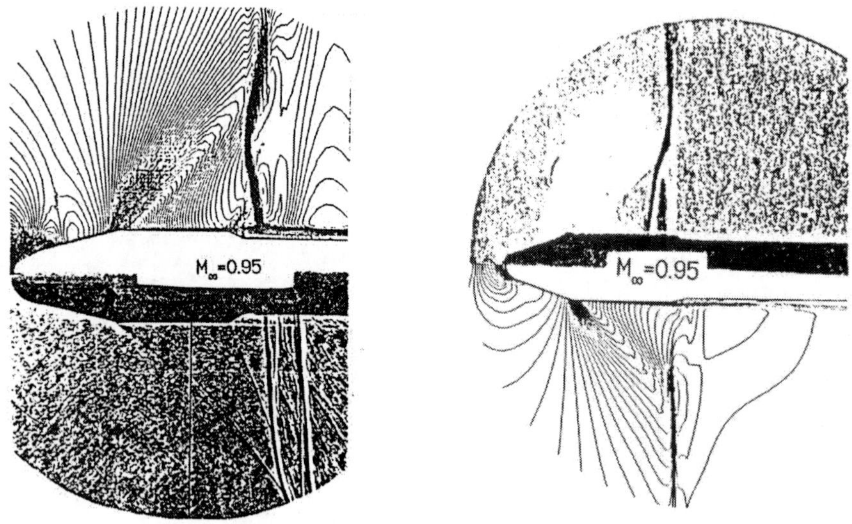

Figure 16. Comparison between density contours and schlieren pictures.

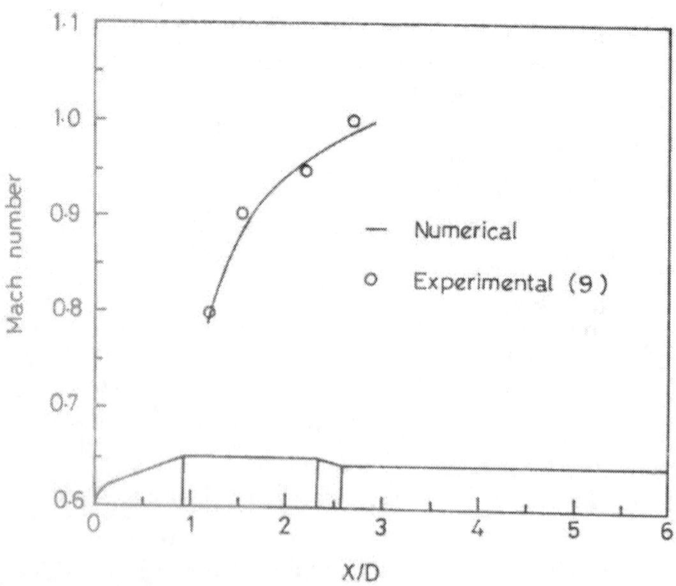

Figure 17. Shock location at various Mach number.

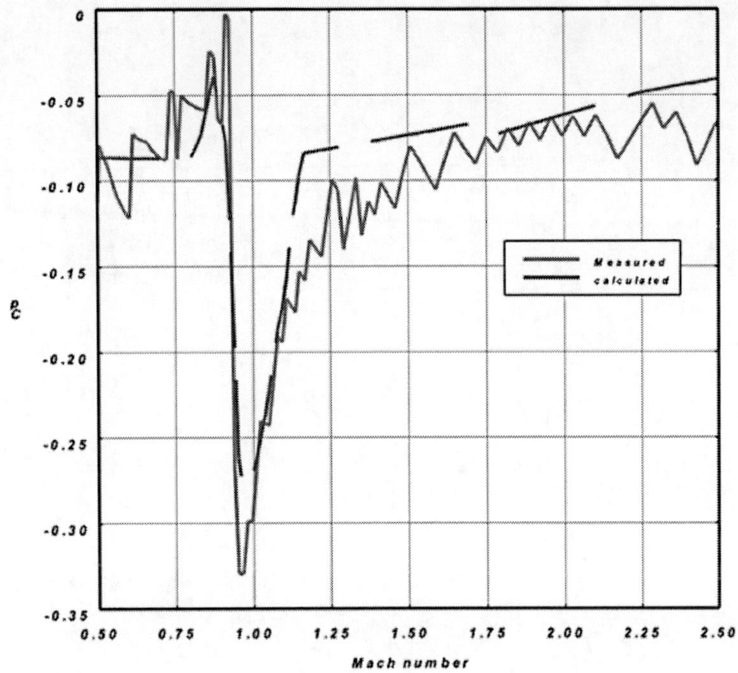

Figure 18. Variation of pressure coefficient on the heat shield.

In numerical studies [29], a local flow separation is observed in the boat-tail region and is found to be a function of Mach number. A schematic sketch of the separated flow in the boat-tail region at the transonic and supersonic Mach numbers, which are reconstructed from the velocity vector and density contour plots, is shown in Figure 19. For the transonic case, a shear layer is formed, and it contains a recirculating region as shown in Figure 19(a). The downstream boundary layer is found to be thick, which is nearly the boat-tail height. Figure 19(b) shows flowfield features for the supersonic case. The flow turns downstream of the boat-tail through an expansion fan with a resulting mixing layer that separates the supersonic fluid above the subsonic recirculating region below it. The pressure in the recirculating region is generally uniform and below that of freestream pressure. As the mixing layer impinges on the downstream wall, the flow turns back toward the freestream direction and a recompression shock wave is typically formed. The flowfield visualization of the terminal shock and the separated region helps in a systematic understanding of flow structure under various freestream Mach numbers. In the boat-tail region, a local separation results, due to sharp discontinuity in the longitudinal curvature. The separated flow in the boat-tail region is observed for both heat-shield configurations. The flow reattachment length is normalized by the boat tail height, where X_c is the location of the boat-tail shoulder and X_r is the reattachment point as depicted in the insert of Figure 20.

The prediction of Xr was chosen as the point where the axial component of velocity along the downstream wall changed from negative to positive. Figure 20 shows the comparison between measured normalized separation lengths with the computed value. A reasonably good agreement is seen between them. The separation length is function of Mach number. The extent of predicted separation reaches its maximum value when the freestream Mach number reaches unity. Assessment of the separation region and knowledge of the surface pressure distribution is required for acoustic insulation design.

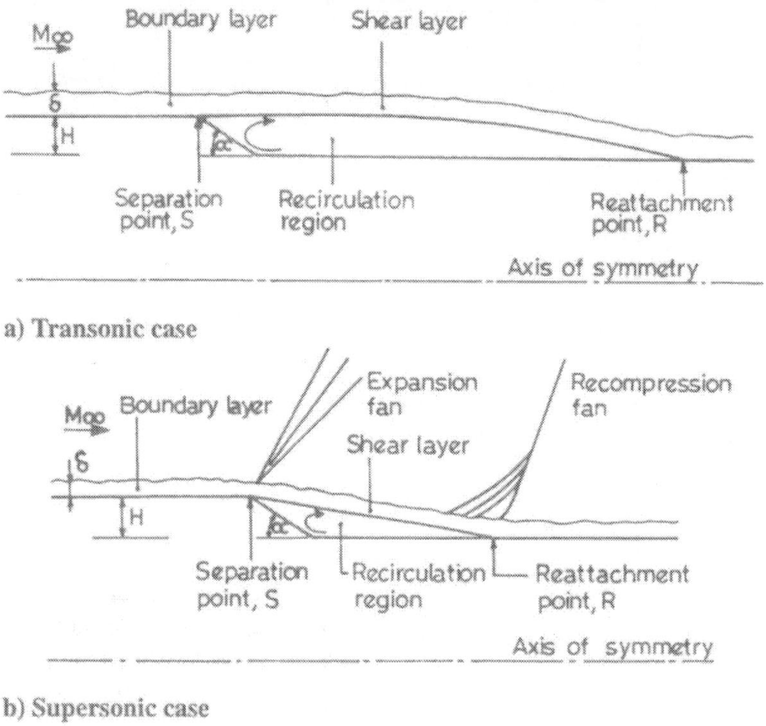

Figure 19. High-speed flow over boat-tail region of payload.

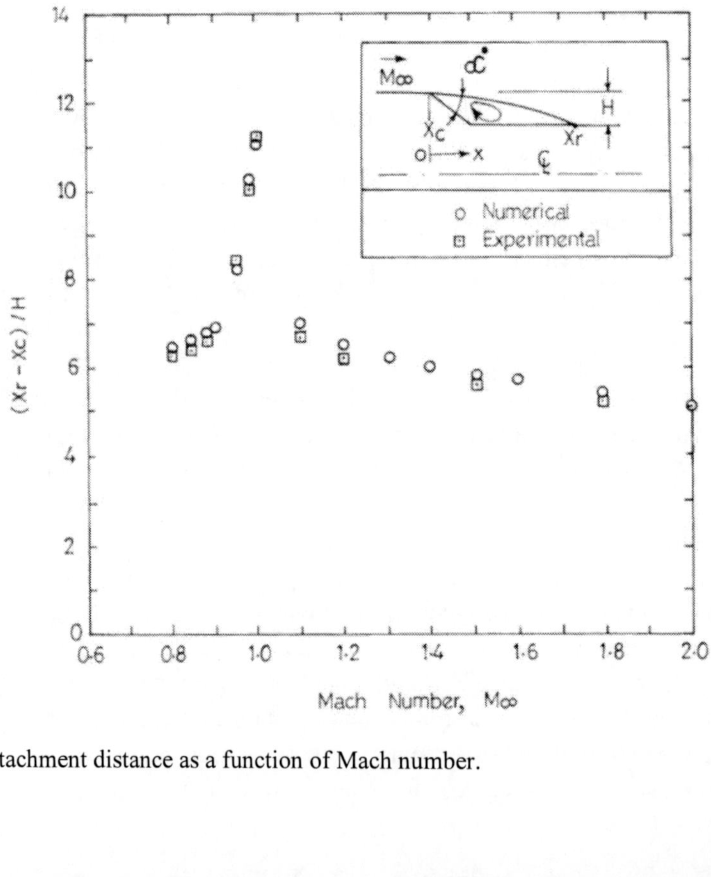

Figure 20. Reattachment distance as a function of Mach number.

3. UNSTEADY FLOW OVER PAYLOAD FAIRING

The occurrence of buffeting during atmospheric ascent depends primarily on the shape of the heat shield configuration. To eliminate this problem, the NASA guidelines [24] recommended certain values of the cone angle, cone and cylinder lengths, and boat-tail angle of the payload shroud. The buffet loads must be considered in the design of the space-vehicle structure. The pressure fluctuations which originated at the location of the normal shock wave within regions of separation can cause buffeting of the space vehicle. The buffet pressures are configuration dependent. Using Gaussian distribution, the appropriate unsteady pressure data are utilized to calculate the sound pressure level and corresponding frequencies. The time-dependent calculation reveals that around the cone-cylinder junction near the shock the SPL values are high and coupled with the low frequencies. The vehicle should be designed to minimize buffeting through the use of favorable configurations. The occurrence of buffeting of vehicle during atmospheric ascent depends primarily on the shape of the vehicle [24].

Coe [30] has investigated pressure fluctuations over hammer head pay load shroud. The buffet pressures are configuration dependent [31]. Self-sustained oscillation [32] of launch vehicles can result during transonic flight if structural responses couple with alternating flow separation forces at the payload fairing cone-cylinder junction. Pressure and heat flux fluctuations due to shock-boundary layer interaction show that the fluctuations in heat flux quantitatively the same feature as those of wall pressure but different quantitatively [33].

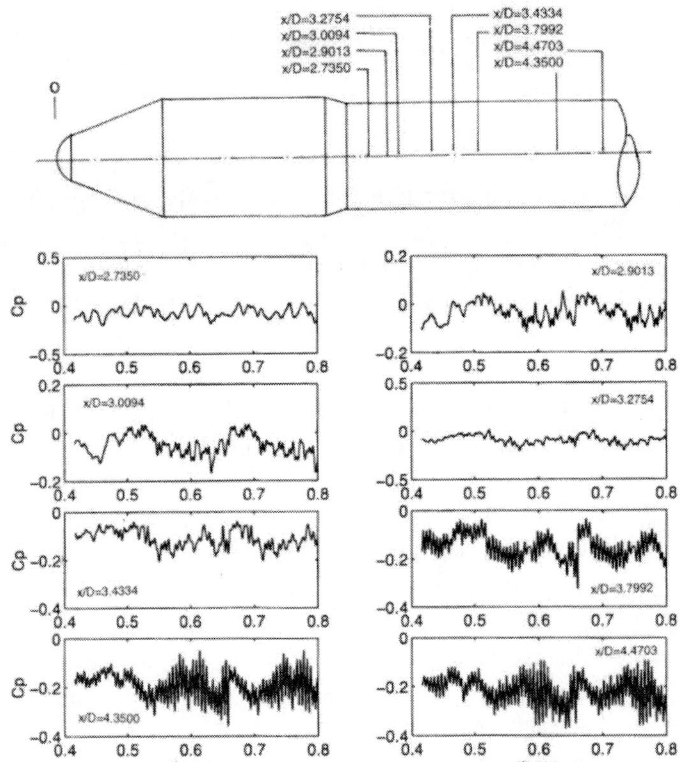

Figure 21. Surface pressure fluctuations for heat shield model M1.

Table 4. Overall Sound Pressure Level on Payload Fairing

Mach number	OASPL, dB
0.80	158.6
0.85	158.0
0.90	157.6
0.95	153.7
1.00	158.2
1.07	158.4

Using the unsteady numerical simulation [34] of the RANS, the unsteady flow field analysis is carried out for the heat shield M1 and M2. Figure 21 shows the surface pressure fluctuations for the model M1. A cyclic nature of the surface pressure variation occurs at x/D = 2.4611, and with a time period of about 1.5 s for one cycle. A set of histograms of pressure coefficient for the transonic Mach number range for M2 heat shield is shown in Figure 22. A spectral analysis was carried out on the computed pressure data for all possible modes of oscillations using fast Fourier transform FFT of MATLAB [35]. The FFT converts the pressure histories from time domain into the frequency domain. The pressure values have been converted from Pascal to decibel (dB) of surface pressure levels. The surface pressure levels are computed in terms of the rms pressure reference at 20μPa. Table 4 gives Overall Sound Pressure Level (OASPL) for various values of freestream Mach number. A cyclic behavior of the pressure coefficient is observed in the vicinity of the separation and reattachment points. From the spectrum analysis low frequency pressure fluctuations and sound pressure levels are found at freestream Mach number 0.95, whereas high frequency fluctuations are observed at supersonic Mach number 1.20 as presented in Ref. 36.

The inclusion of body flexibility in dynamic loads and stability analyses requires a detailed description of the normal force distribution on body by integrating the running load curve over the length of the body and dividing by a reference area, a total normal force coefficient slope can be obtained. An empirical method based on experimental results for rapid determining distributed aerodynamic coefficient configuration has been developed. It represents a compilation and correlation of available experimental results into a set of design curves that yield aerodynamic distribution for various geometric components used in launch vehicle design. The tangent-cone method [37] is useful for quick calculation of the pressure distribution over slender body of revolution at angle of attack. The stability derivatives are in error when they are determined by the tangent-cone method, however, predicts the correct trend of the derivative due to nose blunting but over estimates the magnitude. An experimental investigation [38] and [39] has been made to determine the normal force coefficient slopes, centre of pressure locations, and fore body drag coefficients for several cones with different nose bluntness ratios over a Mach number range from 0.5 to 5.0. The local normal load distribution, CN_{α} normal force coefficient and centre of pressure X_{CP} are calculated over the sounding rocket using tangent-cone method in-conjunction with numerical simulation of the compressible axisymmetric Euler equations. The Mach contours plot for freestream Mach numbers 0.9 – 1.5 are shown for typical sounding rocket in Ref. [40]. Table 5 gives the comparison between the experimental data with aerodynamic coefficient obtained using the hybrid method. The hybrid method is simple, computationally efficient for getting preliminary aerodynamic coefficient at supersonic Mach number at moderate angle of attack.

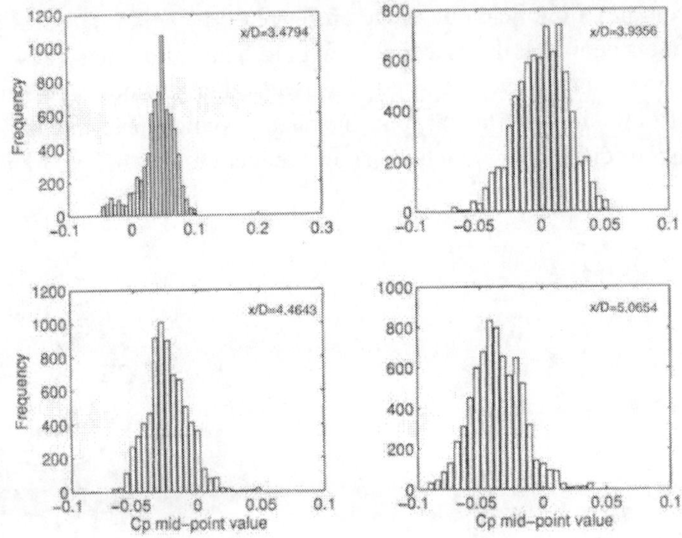

Figure 22. Histogram of pressure coefficient (M2 heat shield).

Table 5. Aerodynamic coefficient for sounding rocket

Mach Number	Experimental		Calculated	
	CN_α	X_{CP}	CN_α	X_{CP}
2.5	0.058	8.60	0.051	8.50
3.0	0.60	8.70	0.059	8.40

4. Flow Field Study of the Forward Facing Spike

High-speed vehicles are designed to withstand severe aerodynamic heating conditions. Such vehicles include hypervelocity projectiles, reentry payload modules, and hypersonic aircraft. Maximum heating and the consequent potential for material erosion are typical problems associated in the nose region of the blunt body. Therefore, there exists a need to develop active or passive technique to reduce the heating rates in the vicinity of the nose tip region of the vehicle. Heat flux can be reduced by attaching a forward facing spike on the blunt nose of the vehicle. The spike is also an effective way to reduce the aerodynamic drag due to the reduced dynamic pressure in the separated flow region.

A forward facing spike attached to a heat shield creates a region of flow separation over the nose that reduces the aerodynamic drag [41] due to the reducing dynamic pressure in the separated region at supersonic and hypersonic speeds. A summary of experimental investigations up to 1966 has been reported in Ref. [42] for axially symmetric models with nose shapes of hemispheres, flat surfaces, hemispherical flats, ogives and tapered configurations, and cylindrical after bodies in the freestream Mach number range of 1.75 ≤ $M_\infty \geq 14.0$. A complete flow field over the heat shield with forward facing spike [43] is presented in Figure 23 for $M_\infty = 1.7$ and compared with schlieren picture. The density contour picture show a complex flow field due to the presence of the spike. The schlieren picture and density contour plots depict all the essential flow field features of the supersonic flow such as formation of the bow shock wave over the spike, reattachment shock wave and expansion fan over the shoulder of the forward facing spike attached to typical heat shield of launch vehicle. Heat transfer study of the spiked blunt body shows reduction of wall heat transfer [44].

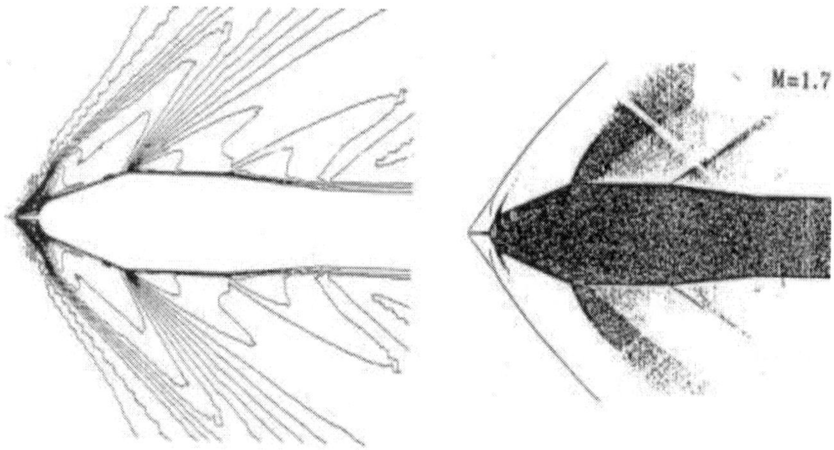

Figure 23. Comparison between density contour and schlieren picture for a forward facing spike attached to heat shield.

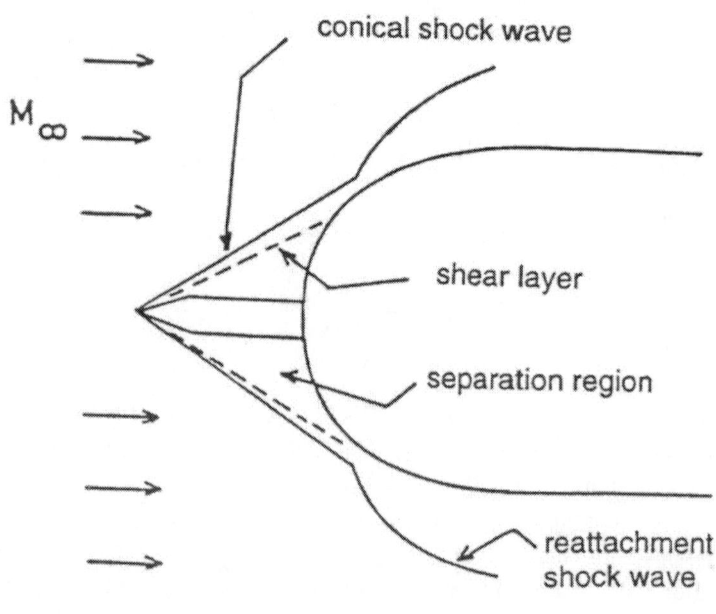

Figure 24(a). Schematic sketch of flow field over spiked blunt body.

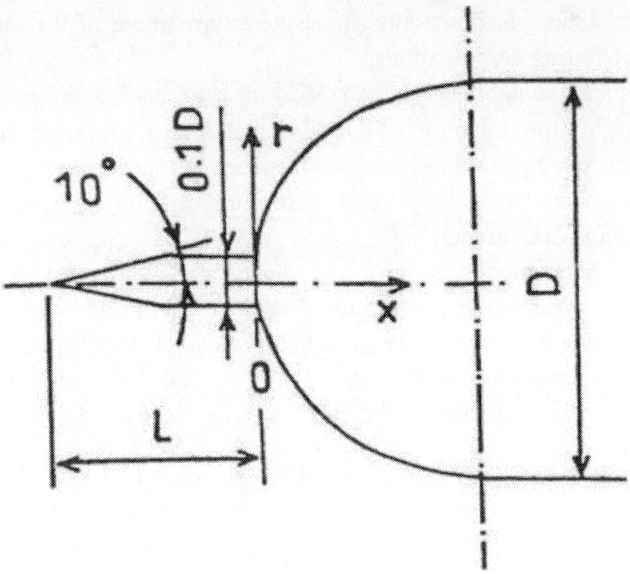

Figure 24 (b). Dimension of the spiked blunt body.

Fundamentally the flowfield instability in was observed in earlier 1950's, when drag reducing spikes considered to be employed on ballistic missiles. The features of the supersonic flow field over a conical spike attached to hemispherical body can be delineated through the experimental investigation. It is characterized by a conical shock wave from the tip of the spike, a reattachment shock wave on the blunt body, and a separated flow region ahead of the blunt body. A schematic of the flow field over the spiked blunt body is shown in Figure 24 (a). This flow field has also been numerically investigated and physical aspects of the flow field have been found. The dimensions of the spiked blunt body are shown in Figure 24(b). The model is axisymmetric, the main body has a hemisphere-cylinder nose, and diameter D is 7.62×10^{-2} m. the spike consists of a conical part and a cylindrical part. The angle of the spike's cone is 10°, and the diameter of the cylinder of the spike is 0.1 D. Spike lengths L of 0.5D, 1.0D, and 2.0D are considered in the numerical simulation. Figure 25 shows the computed flow field for the spike of L/D = 0.5, 1.0, and 2.0 at $M_{\infty} = 6.8$. Flow patterns for L/D = 1.0 and 2.0 are similar to that of L/D = 0.5. However, when the spike length is large, the angle of the oblique shock wave from the spike tip decreases, and flow separation occurs slightly downstream. The angle of the conical shock wave depends on the angle of the shear layer, which gives an effective body shape to the outer freestream. The computed flowfield shows agreement with the schlieren photographs taken in the experiment by Crawford [45]. According to these computed flow fields, the angle of the conical shock wave is much larger than that of a single cone of 10^{0}. This difference may be attributed to the recirculation region acting as if it were a body boundary. This feature effectively causes the reduction in the drag.

The occurrence of buzzing might be critical because it brings about severe pressure oscillations on the body attached with the spike. Disk on the shaft reduce stable recirculation regions by dividing single separation flow into several critical cavity flows.

The aerospike [46] substantially reduces missile forebody drag by creating a low dynamic pressure region of separated flow over the high volume nose fairing. Wind-tunnel tests [47] were conducted to define the fluctuating pressure environment of the telescoping design of the aero spike/nose fairing combinations.

The effects of spike length, shape (aero spike or aero disk), spike nose configuration and angle of attack [48] on the reduction of heat flux and drag is experimentally studied at M_∞ = 7. These experiments suggest that the aerodisk is superior to the aerospike and that the aero disk of appropriate length, diameter, and nose configuration may have the potential for thermal protection and drag reduction.

The hysteresis phenomenon occurring in unsteady flows over an axisymmetric spiked cylinder at Mach 2.21 was numerically simulated by Feszty et al. [49]. One of the advantages of CFD in investigation of the unsteady spiked body flows that it can offer a sequence of flow visualization frames from one particular cycle. This could be difficult to achieve in the experimental due to high frequency of the event, varying from 2 kHz up to 7.5 kHz depending on the Mach number [50].

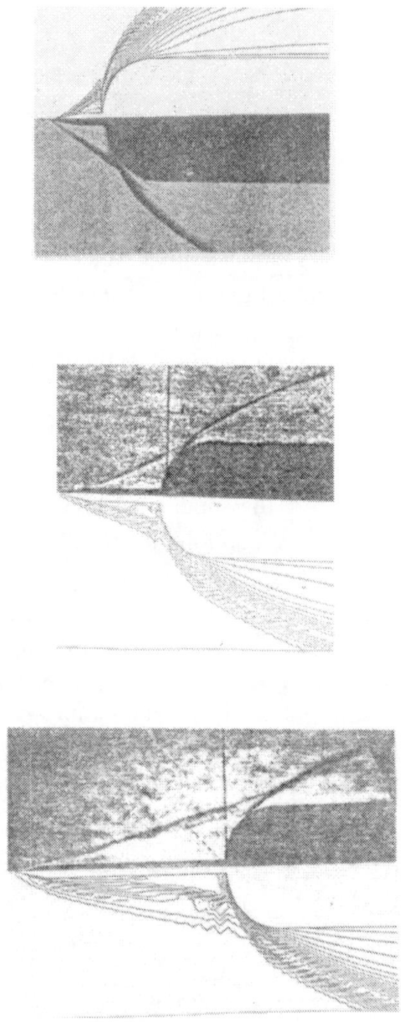

L/D = 0.5

L/D=1.0

L/D=2.0

Figure 25. Comparison between density contour and schlieren picture.

There were few studies directed to reveal the driving mechanisms of unsteady spiked body flows [51]. The phenomenon of self-excited shock oscillations on a spike-tipped body at Mach 3 is investigated experimentally. The shock oscillations are symmetric and asymmetrically depend on spike length [52]. Used the fluid dynamic stability analysis to determine flow frequency regimes where the shear layer would admit disturbance amplification. Supersonic flows around a spiked blunt body [53] are numerically simulated to examine the effects of spike length, Mach number, and angle of attack.

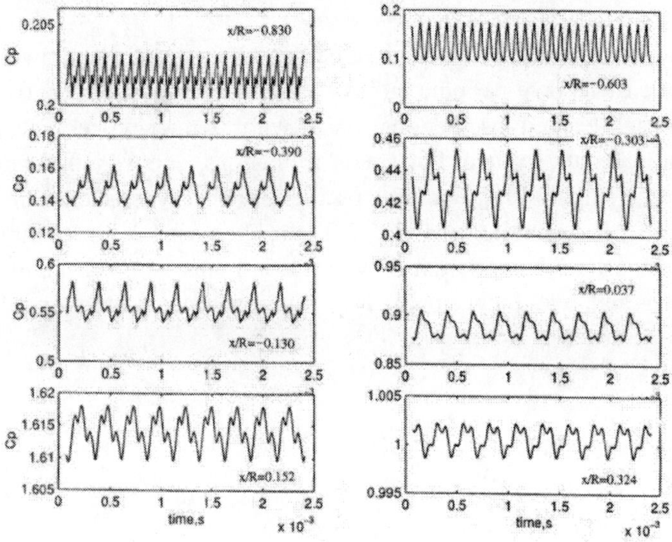

Figure 26. Pressure oscillation over spiked blunt body (L/D=0.5).

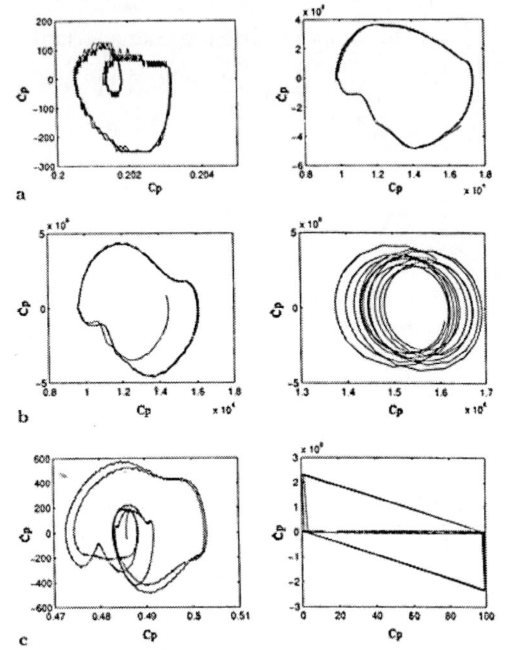

Figure 27. (a) Phase trajectory, L/D = 0.5, (b) Phase trajectory, L/D = 1.0, (c) Phase trajectory, L/D = 2.0.

Computational Fluid Dynamics used to investigate the flowfield characteristics for various types of forward facing spike [54]. The surface pressure oscillations over the spike and phase plots exhibit a behavior analogous to that of the Van der Pol equation [55] for the self-sustained oscillating flow [56].

Numerical results [57] are compared with the experimental data [58] as shown in Figure 25 for L/D = 0.5, 1.0 and 2.0 at $M_\infty = 6.8$. Peak heating for reattchament of separated flow on a spike is investigated for different length to diameter of the spike in Refs. [59] and [60]. Pressure oscillation over the spike for L/D = 0.5 and freestream Mach number of 6.8 is shown in Figure 26.

The self-excited oscillations is self governed or autonomous and drawing its energy from an external source by its own periodic motion. When the motion ceases, the alternating force disappears. The self-excited oscillations occur with frequency very close to the natural frequency of the undamed system. For small oscillations, energy is fed into the system and there is "negative" damping [61]. For large oscillations energy is taken from the system and is therefore damped. To emulate periodic pressure behavior with a differential equation describes the self-sustained oscillation of Van der Pol equation. Assuming that the pressure waves are planar and one-dimensional, the resulting second order, nonlinear differential equation can be written as

$$\frac{d^2p(t)}{dt^2} - \lambda(1 - p(t)^2)\frac{dp(t)}{dt} + p(t) = 0, (l > 1) \tag{18}$$

where p is pressure computed numerically at time t. λ is the aerodynamic damping term. This is having negative value for small oscillations. Figure 27 shows variation of time dependent pressure vs pressure. The phase plane derivative of the pressure coefficient is shown in Figure 27. The phase portrait is analyzed to understand the characteristic of the oscillatory flow. The critical point (0,0) of the Van der Pol equation shows that the critical point is stable and attractive [62].

5. Estimation of Angle of Attack of Launch Vehicle

The accurate measurement of air data is critical to the flight control, guidance and post flight analysis of most modern atmospheric flight vehicles. The Flush Air Data Sensing (FADS) concepts relies on a mathematical model that relates measured fore body surface pressure to the air data state. The current method uses a nonlinear pressure model and a nonlinear algorithm to solve for the air data state from the measured pressures. Five hole probes are used in the space shuttle [63] to estimate flow direction during the ascent period of the flight. The ascent air data system of space shuttle consists of a bi-conic spike, or probe mounted on the tip of the external tank. Out of five probes, four probes are located on the conical region of the spike to measure differential pressure during flight. These pressures probes are sensitive to pitch and yaw angle. A relationship between the wind tunnel calibration and calibrated corrected flight pressure data are required for the estimation of flow direction. Calibration coefficients and parameters are derived using measured pressure data obtained from the wind-tunnel. In the post flight analysis, the measured pressures obtained through the telemetry are used to estimate the pitch, yaw and total angle of attack in conjunction with the wind tunnel data. Neural network method [64] have been developed and implemented for sensing air data obtained from the multi-hole pressure probes. The calibration of multi-holes requires a large data base of flow cone angles. The artificial neural network needs net work architecture, training and optimization for estimating the flow parameters. The calibration of multi-hole probes used in the neural network method needs large number of blow down of wind tunnel. Therefore, the most economical and computationally efficient way is to simulate flow field over the complex aerodynamic configuration using the Computational Fluid Dynamics.

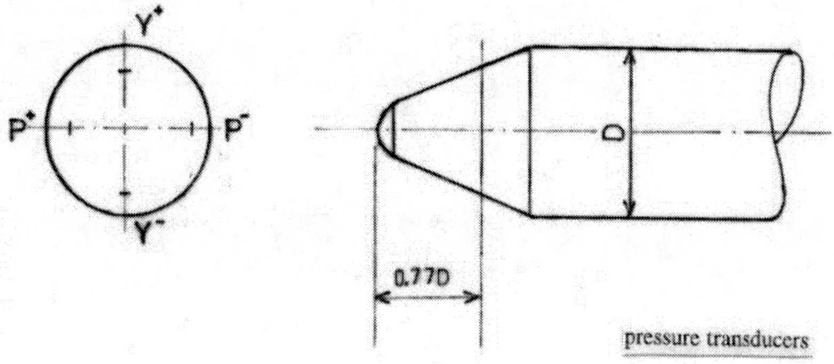

Figure 28. Location of pressure transducers on payload fairing.

Figure 28 shows geometry of the heat shield consists of spherical-cone and cylinder. Probes are mounted at 0.77D from the stagnation point of the heat shield on pitch (P+ and P-) and yaw (Y+ and Y-) planes. D is the diameter of the cylinder region of the heat shield. The time-dependent compressible Euler equations are written in the Cartesian coordinate system as

$$\frac{\partial \mathbf{W}}{\partial t}+\frac{\partial \mathbf{F}}{\partial x}+\frac{\partial \mathbf{G}}{\partial y}+\frac{\partial \mathbf{H}}{\partial z}=0 \qquad (19)$$

where

$$\mathbf{W}=\begin{bmatrix}\rho\\ \rho u\\ \rho v\\ \rho w\\ \rho e\end{bmatrix},\ \mathbf{F}=\begin{bmatrix}\rho u\\ \rho u^2+p\\ \rho u v\\ \rho u w\\ (\rho e+p)u\end{bmatrix},\ \mathbf{G}=\begin{bmatrix}\rho v\\ \rho u v\\ \rho v^2+p\\ \rho v w\\ (\rho e+p)v\end{bmatrix},\ \mathbf{H}=\begin{bmatrix}\rho w\\ \rho u w\\ \rho v w\\ \rho w^2+p\\ (\rho e+p)w\end{bmatrix} \qquad (20)$$

are the W state vector conserved quantities with ρ, u, v , w and e denoting the density, Cartesian velocity components, and the specific total internal energy, respectively, and inviscid flux vectors, E, F and G in the Cartesian coordinate. With the ideal gas assumption, the pressure and total enthalpy can be expressed as

$$\rho e = \frac{p}{(\gamma - 1)} + \frac{1}{2}\rho\,(u^2 + v^2 + w^2) \tag{21}$$

where γ is the ratio of specific heats.

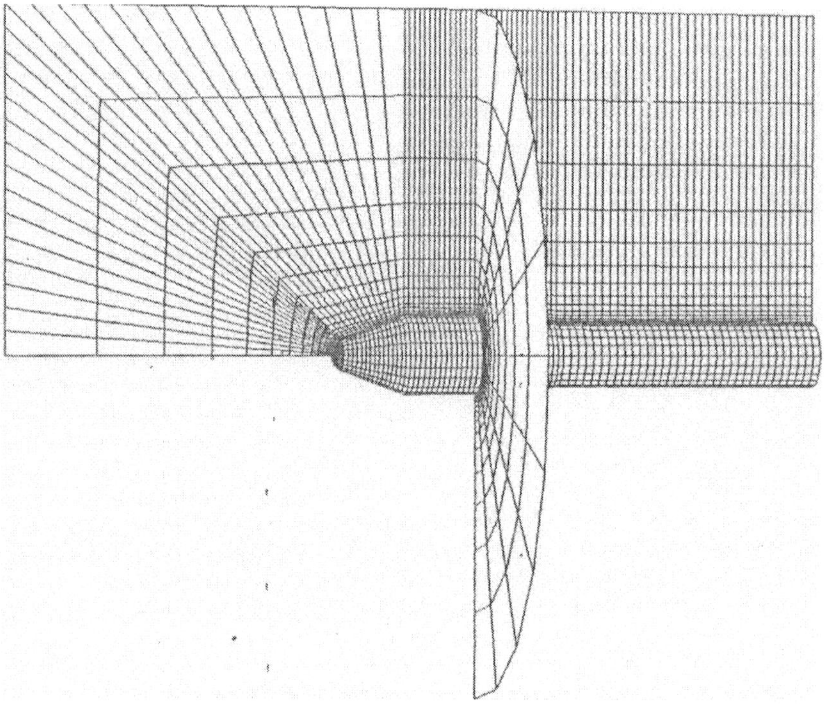

Figure 29. Computational grid over heat shield.

To facilitate the spatial discretization in the numerical scheme, the governing fluid dynamics equation (19) are written in the integral form. Temporal integration is carried out using multistage Runge-Kutta time-stepping scheme. A local time-step is used to achieve steady state solution. Figure 29 depicts grid over the payload fairing. A finite volume scheme in conjunction with multistage Runge-Kutta time stepping is employed to solve the three-dimensional compressible Euler equations. Calibration parameters are derived using the computed differential pressure with small increment of freestream Mach number and angle of attack. The pitch and yaw angles are calculated from the calibration parameters derived from the numerical simulation of the flow field over the heat shield. High degree of linearity in the pitch and yaw angles are seen with respect to the angle of attack. The calibration factors span the freestream Mach number range of 0.5 to 3.0 which covers maximum dynamic pressure and transonic region of the vehicle. The calibration factors are validated with wind tunnel results. Initial results from the ascent air data system are compared with reconstructed trajectory data. The differential pressure transducers are in the range of $\pm 13.788 \times 10^3$ Pa. The digitization of the flight pressure data is 0.41×10^3 Pa for absolute pressure and 0.1096×10^3 Pa for differential pressure.

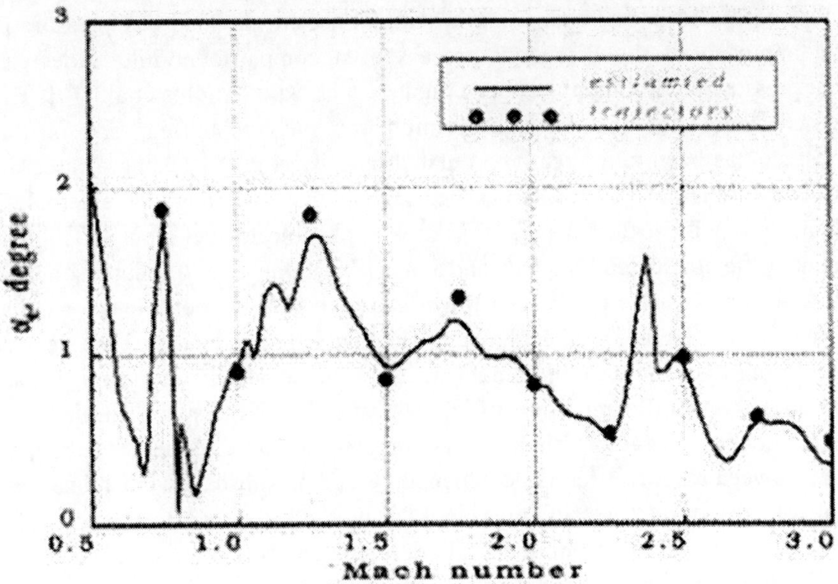

Figure 30. Variation of flow angles with Mach number over heat shield.

Figure 30 shows the variation of flow angles with Mach number over heat shield of a satellite launch vehicle [65]. The results are compared with the reconstructed trajectory data and found in reasonably good agreement. Numerical analysis is employed to compute angle of attack of the launch vehicle in conjunction with the measured pressure during the flight of the satellite launch vehicle. The measurements of the pressure are done in the conical region of the heat shield. The location of the pressure transducers are selected in the conical region of the heat shield because the flow is attached. The present scheme differs with the previously method of the estimation of angle of attack. The prediction of angle of attack is useful parameter for the post flight analysis of satellite launch vehicle.

6. Depressurization of Payload Fairing

During the atmospheric flight, the heat shield fairing of a satellite launch vehicle needs venting in order to prevent abnormal pressure built-up, this can be detrimental to the structure. Important class of aerodynamically induced loads which must be considered in the design of a sounding rocket are those due to the differential pressure occurring across internal bulk heads and across the vehicle outer surface. Improper design of a spacecraft launch vehicle or sounding rocket venting system can result complete mission failure. The static pressure and equivalent dynamic pressures are used in venting analyses and aided in describing the pressure differential predicted along the payload fairing. In order to model the venting process and predict the differential pressure time history reasonably for the design, a compressible flow loss coefficient is essential for the mass flux calculation. International reference guide [66] to space launch gives the maximum differential and rate of change of pressure inside the fairing. The NASA monograph [67] describes the design criteria of compartment venting for space vehicles during ascent and re-entry period. An earlier study was carried out by Mironer and Regan [68] and Murri [69] to determine the venting design criteria for space shuttle payload, using a nominal ascent, payload bay pressure profile. An analytical approach of the discharge process of a compartment into a decreasing time-dependent pressure environment has been published by Sanz-Andres et al. [70]. The effective discharge coefficient for flow through the multi-row payload fairing vents on the Titan IV launch vehicle has been numerically simulated by Huseman [71] using a three-dimensional analysis. The discharge coefficient has been obtained for a range of external flow Mach number and internal-to-external pressure ratios applicable to the Titan IV flight trajectory. Experiments were conducted by Johns and Jones [72] in the 8ft x 6 ft supersonic wind tunnel of the NASA Lewis Centre to find out the effective discharge coefficient for the application of the Titan/Centaur launch vehicle. Fay and Hengel [73] have analyzed the flow through the vent connecting the multi-compartment using a quasi-steady isentropic equation with empirical discharge coefficient. A post flight analysis [74] of the space shuttle has shown that the differences between the pre-flight prediction and measured values were primarily due to difference between external pressures, which are based on subscale wind-tunnel test data, and the actual vehicle external pressure measured during the flight. Most of the predictions of the discharge coefficient are based on the flight derived vent port pressure coefficients since the wind tunnel does not adequately define the orbiter ascent pressure environment. An inverse venting problem [75] for estimation of the discharge coefficient in a heat shield fairing carried out in conjunction with the measured internal pressure. An orifice exits between the zero-dimensional models and the full two-dimensional/three-dimensional models of a depressurization loads. To combine the accuracy and conservation properties of the full numerical simulation methods with the efficiency of the simplified boundary conditions, a new reduced-order model of a multidimensional venting from a heat shield of a typical launch vehicle is proposed. In contrast to the methods available in the literature, the new approach uses a reduced-order model to approximate a two-dimensional or three-dimensional venting problem. The multi-dimensional depressurization is simulated by the solution of the time-dependent Euler equations similar to those used for the quasi-one-dimensional nozzle problem. In contrast to the zero-dimensional models, this approach provides conservation of mass, momentum, and energy. Numerical results demonstrating the accuracy of the new-reduced-order model and its limitations are presented. The low-dimensional depressurization model of venting has a several advantages. First, this approach is fully conservative and provides conservation of mass, momentum, and energy. Second, the new quasi-one-dimensional model is computationally much more efficient compared with two-dimensional or three-dimensional numerical simulations of the orifice flow. Third, the reduced-order model retains some important features of the realistic venting model, such as the area of the vent holes, volume of the compartment. These quantities of the new model and its ability to account for the compressibility effects inherent in the venting analysis make it an efficient tool for quantitative study of the venting performance as a function of trajectory. The proposed reduced-order model is based on the assumption that the multidimensional and viscous effects inside the compartment away from the orifice can be neglected.

The reduced-order model of multidimensional payload venting is proposed for numerical simulation of simplified zero-order model [76]. The multidimensional venting analysis is simulated by the solution of the time-dependent compressible-quasi-one-dimensional Euler equations. The governing quasi-one-dimensional equations can be written in conservation law form as follows:

$$\frac{\partial \mathbf{W}}{\partial t}+\frac{\partial \mathbf{F}}{\partial x}+\mathbf{S}=0 \tag{22}$$

where

$$\mathbf{W}=A\begin{bmatrix}\rho\\ \rho u\\ \rho e\end{bmatrix},\quad \mathbf{F}=A\begin{bmatrix}\rho u\\ \rho u^2+p\\ (\rho e+p)u\end{bmatrix},\quad \mathbf{S}=-\begin{bmatrix}0\\ p\dfrac{\partial A}{\partial x}\\ 0\end{bmatrix} \tag{23}$$

where W is conserve vector and F is flux vector. A is the cross-sectional area of the quasi-one-dimensional depressurization equation for the venting. It is considered that A is a continuously differentiable function that is independent of time. Therefore, the A is function of axial distance, x and can be expressed as A = A(x). The variablesρ, u, p, e and H are the density, velocity, pressure, total specific energy and total enthalpy, respectively. S is source term due to the area variables. The equation of state for perfect gas is

$$p=(\gamma-1)\rho(e-\frac{1}{2}u^2) \tag{24}$$

where γ is the ratio of specific heats for air.

The numerical method uses a finite volume discretization scheme. The spatial discretization reduces the governing equations to semi-discrete ordinary differential equations. The integration is carried out using a three-stage Runge-Kutta time-marching method. It is important to mention here that the ambient conditions of the vent depend on the trajectory of the satellite launch vehicle. The differential pressure can be computed as

$$\Delta p=(p_c-p_e) \tag{25}$$

where p_c and p_e are heat shield compartment and ambient pressure respectively. The p_c is computed by solving Eq. (23). The differential pressure is an important parameter for the structural load and proper functioning of the electronic components inside the heat shield. The present numerical simulation is used to compute pressure inside the heat shield during the ascent phase of the launch vehicle. The compartment pressure is calculated using following expression [70]

$$\frac{d\rho_0}{dt}=-C_D\frac{A_H}{V_C}\left(\frac{2\gamma}{\gamma-1}\right)^{\frac{1}{2}}\left[\left(\frac{p_i}{p_e}\right)\left(\frac{\rho_0}{\rho_i}\right)^{n}\right]\left[\frac{p_i}{\rho_i^{n}}\right]^{\frac{1}{2}}\left(\rho_0^{\left(\frac{n+1}{2}\right)}\right)\left[1-\left\{\left(\frac{p_i}{p_e}\right)\left(\frac{\rho_0}{\rho_i}\right)\right\}^{\frac{\gamma-1}{\gamma}}\right]^{\frac{1}{2}} \tag{26}$$

where n is polytropic coefficient. Subscripts e, i, and o are condition inside the fairing, initial condition and inside respectively. A_H is the area of the vent hole, V_C is volume of the compartment, ρ is density, C_D is the discharge coefficient, p is pressure, γ is the ratio of specific heats. The variation of discharge coefficient, C_D is shown as function of flight time in Figure 31 for the fairing of M1 and M2 launch vehicle. The discharge coefficient is defined here as the actual mass flow rate divided by the ideal mass flow rate through the vent hole.

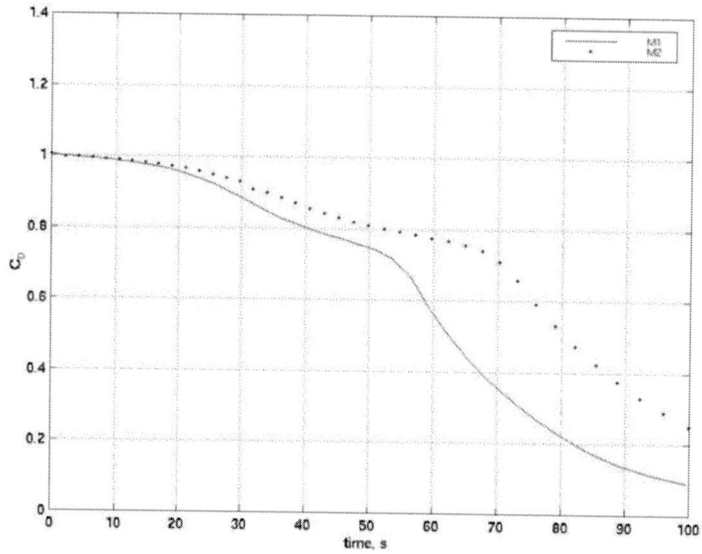

Figure 31. Variation of discharge coefficient as a function of flight time.

The venting analysis leading up to the first Titan IV launch [77] for complete description of the dynamic pressures were combined in determining the skin-pressure differential for the design of pay load fairing. The differential pressure depends mainly on the geometry of the venting holes, the effective volume of air to be evacuated and the trajectory of the launch vehicle.

7. Flow Field Simulation of the Reentry Capsules

The primary design consideration of reentry capsules requires large spherical nose radius of their forebody that gives high aerodynamic drag and a short body length for reducing the total structure weight and the ballistic coefficient. The forebody shape of reentry capsules can be selected either employing a spherical cap, or a combination of spherical nose with cone, or a spherical blunt cone/flare configuration. To ensure the deployment of parachute in the transonic region, the flow field past the capsule must be known at supersonic speeds. The flow field in the wake region of a reentry capsule is complex due to the expansion at the shoulder and the base-shell.

A large number of computational fluid dynamics simulations [78-83] have been performed for aerobraking and reentry capsules. Allen and Cheng [81] have carried out the numerical solution of Navier-Stokes equations in the near wake region of the reentry module, which confirms the mechanism of flow separation as, observed experimentally [83]. Base drag represents the loss in recovery of pressure over the base of the capsule [84]. The supersonic and hypersonic laminar flow over a slender cone has been numerically calculated by Tai and Kao [85]. A summary of developments relating to the base pressure prediction is reported in the review paper of Lamb and Oberkampf [86]. An aerodynamic analysis of the Commercial Experiment Transport (COMET) reentry capsule carried out by Wood et al. [87] by solving the laminar thin layer Navier-Stokes equations flow solver LAURA. The flow field past blunt and short reentry capsule has been analyzed in order to understand the mechanism of the instability at supersonic speeds due to decay of base pressure [88]. Yamamoto et al. [89] have computed flow field over the OREX reentry module in conjunction with the in-depth thermal analysis of thermal protection system and results were compared with the flight data. Tam [90] has used LUSGS implicit scheme for flow computation over On-Axis Biconic and Aeroassist Flight Experiment (AFE) reentry vehicles. Liever et al. [91] solved the flow field over Beagle2 reentry capsule. The flow field and the heat flux computation over the Mars pathfinder vehicle has been numerically carried out by Haas et al. [92] along with fore body and wake flow structure during atmospheric entry of the spacecraft.

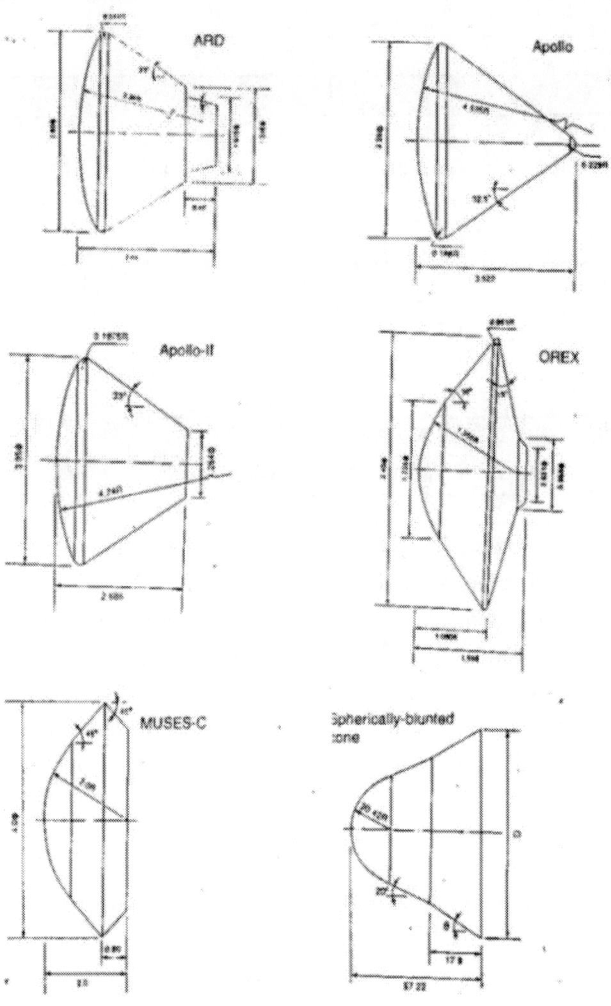

Figure 32. Geometrical details of various reentry configurations.

Table 6. (a) Geometrical parameters of reentry configurations

Capsule	Overall length, L	Semi-cone angle, α_A deg	Back-shell angle, α_B deg	Spherical radius, R_N	Frontal diameter, D	Corner radius, R_C
ARD	2.04	–	33.0	3.36	2.80	0.014
Apollo	3.522	–	32.5	4.59	3.95	0.186
Apollo-II	2.686	–	33.0	4.74	3.95	0.197
MUSES-C	2.0	45.0	45.0	2.0	4.0	--
OREX	1.508	50.0	75.0	1.35	3.40	0.001

Table 6. (b) Geometrical parameters of spherical blunt cone-cylinder configurations

Base diameter, D mm	Semi-cone angle, θ in deg
73.91	25
77.89	30
82.29	35

Table 7. Trajectory points and initial conditions

M_∞	U_∞, m/s	P_∞,Pa	T_∞, K
1.2	351	4519	210
1.4	409	3952	213
2.0	596	2891	219
3.0	903	2073	224
5.0	1532	1238	232
6.0	1890	1064	234

The above mentioned reentry geometrical configurations reveal that the forebody shape of reentry capsules can be classified either using as a spherical cap as in the case of Apollo and ARD, or combination of the spherical nose with cone as in the case of OREX and MUSES, or a spherical blunt-cone/flare configuration. Figure 32 and Table 6(a) shows the geometrical details of the various reentry capsules. The semi-cone angle of the spherically-blunted cone flare is given in Table 6(b). The significant flow features over the reentry capsule are described by the following. In the forebody region, the fluid decelerates through the bow shock wave depending upon the cruise speed and altitude. At the shoulder of the capsule, the flow turns and expands rapidly, that the boundary layer detaches, forming a free shear layer that separates the inner recirculating flow region behind the base from the outer flow field. The latter is recompressed and turned back to freestream direction, first by so-called lip shock, and further downstream by recompression shock. At the end of recirculating flow past the neck, the shear layer develops in the wake trail. A complex inviscid wave structure often includes a lip shock wave (associates with the corner expansion) and wake trail (adjacent to the shear layer confluence). The corner expansion process is a modified

Prandtl-Mayer pattern distorted by the presence of the approaching boundary layer. Figure 33 shows schematic feature of the flow field over OREX and spherical blunt cone/flare capsules.

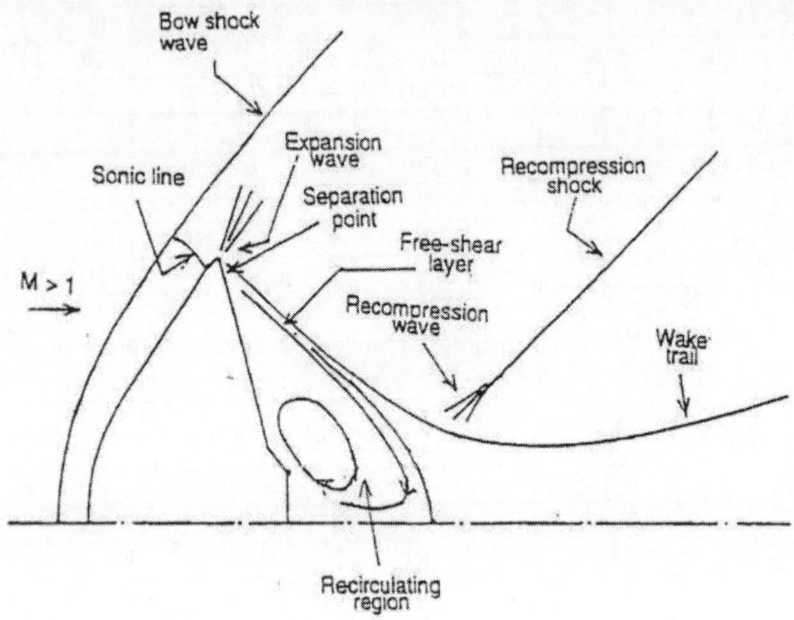

(a) Flowfield over OREX

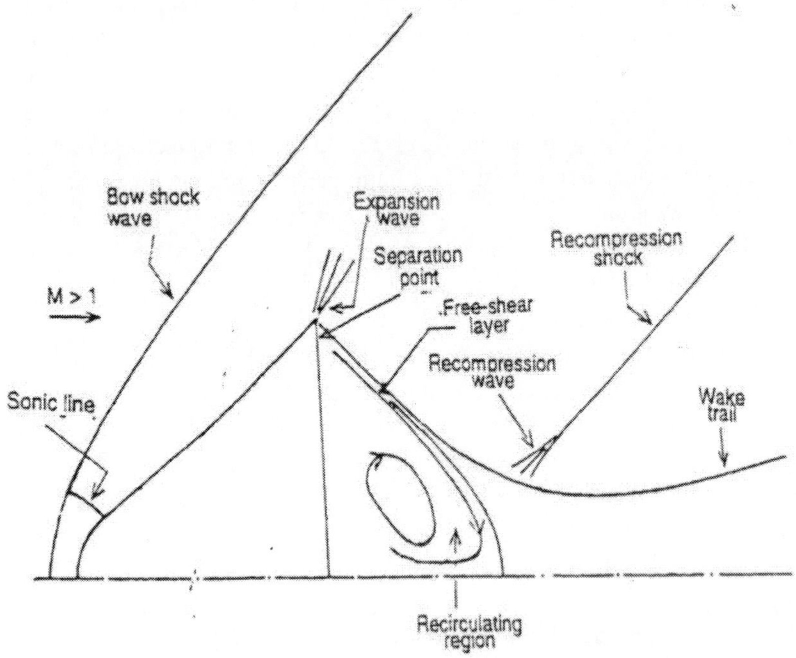

(b) Flowfield over spherical blunt cone module

Figure 33. Representation of flow features on reentry capsules at supersonic-speeds.

Table 8(a). Fore body aerodynamic drag

Capsules	M_∞=1.2	M_∞=1.4	M_∞=2.0	M_∞=3.0	M_∞=5.0
	C_D				
ARD	1.41	–	1.43	1.40	1.38
Apollo	1.30	–	1.50	1.43	1.40
Apollo – II	1.32	–	1.47	1.43	1.05
OREX (with smooth shoulder)	1.79	1.77	1.50	1.30	1.16
OREX (with sharp shoulder)	1.80	1.66	1.36	1.17	1.05
MUSSES – C	1.27	–	1.24	1.36	1.38
Beagle – 2	1.58	1.59	1.54	1.47	1.42

Table 8(b). Forebody drag for spherical blunt cone-cylinder modules

Mach	C_D			
	Experimental	Numerical		
	Θ = 25 deg	θ = 25 deg	θ = 30 deg	θ = 35 deg
1.2	0.43	0.42	0.51	0.58
2.0	0.42	0.41	0.49	0.56
3.0	0.39	0.40	0.46	0.51
6.0	0.37	0.38	0.42	0.47

The sonic line is located on the OREX-type configuration on the shoulder whereas in the case of spherical blunt cone/flare's module, the location of the sonic line is at the junction of the spherical blunt cone. Thus, it is seen that the flow field over the reentry capsule needs a high drag with good static stability margin which leads to the selection of an axisymmetric shape of large angle sphere-cone combination. Table 7 shows the trajectory points and initial conditions for the numerical simulation. The numerical simulation is carried out over the various reentry configurations [93] and [94]. Table 8(a) shows computed value of the forebody aerodynamic drag CD for ARD, Apollo, Apollo II, MUSSES-C, and OREX. Forebody aerodynamic drag over the spherical blunt-cone-cylinder modules are given in Table 8(b). CD increased as the nose radius is increased., as expected, the increase is relatively small. High aerodynamic drag is seen in the ARD- and the OREX-type reentry module as compared with the spherical blunt configurations. A good agreement is found between the calculated values of forebody aerodynamic drag coefficient of the spherical blunted cone-fare reentry configuration with experimental data [95].

The Beagle-2 is analyzed here in order to demonstrate a single block grid arrangement. The dimensional detail of the Beagle-2 capsule is an axisymmetric design with a spherical blunt nose diameter D = 2.8 m, the forebody shape consists of R_N = 1.35 m, a half-angle cone of α_N = 50 deg, L = 1.508 m, and R_C = 0.01 m. The aft body is having a α_B = 15 deg, half-angle cone relative to the plane of symmetry. The body oriented grids are generated using a finite element method [96] in conjunction with homotopy scheme [97]. The reentry capsule is defined by a number of grid points in the cylindrical coordinate system. Using these surface points as the reference nodes, the normal coordinate is then described by exponentially structured field points, $(x_{i,j}, r_{i,j})$ extending outwards up to an outer computational boundary.

Sufficient grid points are allotted in the shoulder region of the capsule. The stretching of grid points in the normal direction is obtained using the following expression

$$x_{i,j} = x_{i,0}\left[\frac{e^{\frac{(j-1)\beta}{nr-1}}-1}{e^{\beta}-1}\right] + x_{i,w}\left[\frac{e^{\frac{(j-1)\beta}{nr-1}}-1}{e^{\beta}-1}\right] \quad (27)$$

$$r_{i,j} = r_{i,0}\left[\frac{e^{\frac{(j-1)\beta}{nr-1}}-1}{e^{\beta}-1}\right] + r_{i,w}\left[\frac{e^{\frac{(j-1)\beta}{nr-1}}-1}{e^{\beta}-1}\right]$$

$$i = 1, 2, 3, \ldots\ldots, nx$$

$$j = 1, 2, 3, \ldots\ldots, nr$$

where $r_{i,w}$ and $r_{i,0}$ are wall and outer surface points, respectively, and β is a stretching factor. nx and nr are the total number of grid points in the x and r directions, respectively. These stretched grids are generated in an orderly manner. A close-up view of the computational grid over Beagle-2 is shown in figure 34. Computed Mach contour plot over the Beagle-2 is depicted in figure 35 for $M_\infty = 1.4$. Characteristic features of the flow field around the blunt body at supersonic speeds, such as bow shock wave ahead of the capsule, the wake, and the recompression shock waves emanating from the neck point, are observed in the Mach contour plot. The bow shock wave follows the body contour and the forebody is entirely subsonic upto the corner tangency point of the Beagle-2 where the sonic line is located. The effects of the sonic line on the aerodynamic design parameters are investigated [94] and [95]. A low pressure is formed immediately downstream of the base which is characterized by a low speed recirculating flow region which can be attributed to fill up growing space between the shock wave and the body.

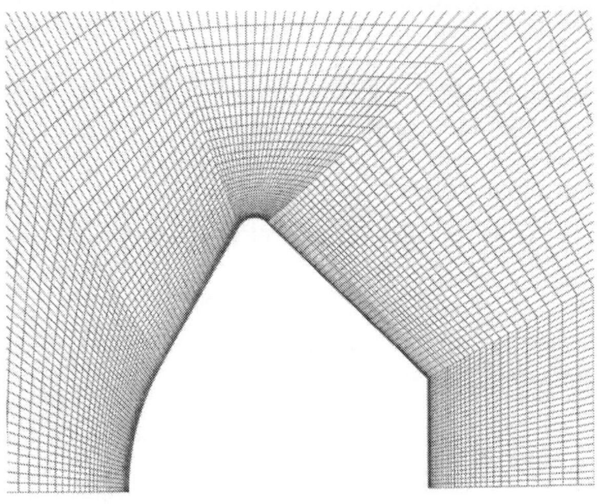

Figure 34. Enlarged view of grid over Beagle reentry capsule.

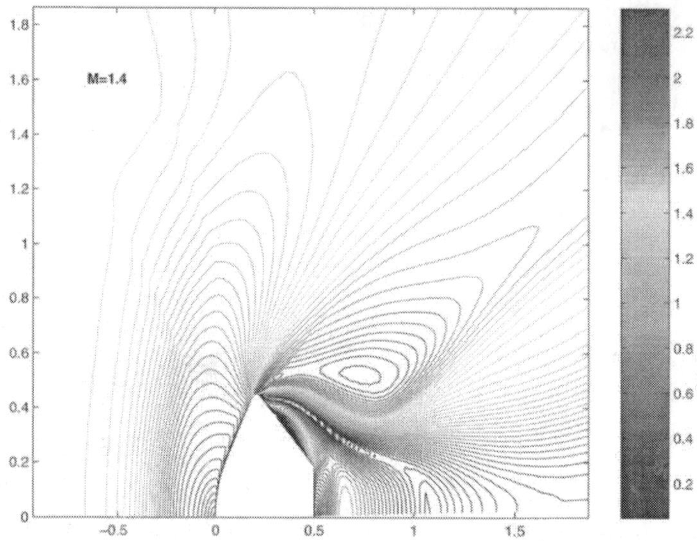

Figure 35. Mach contours over Beagle reentry module.

Figure 36 shows the pressure coefficient $[C_p = \{(p/p_\infty) - 1\}/(2\gamma M_\infty^2)]$ variation along the surface over OREX and spherical blunted-cone flare reentry capsules and freestream Mach number range of 1.2 – 6.0. The s/D = 0 location is the stagnation point, where s is the distance measured along the surface from the stagnation point D is the maximum diameter of the capsule. The pressure coefficient on the spherical cap of the capsule decreases gradually for a given M_∞. The pressure coefficient C_p falls on the sphere-cone junction and remains constant over the cone. In the case of the OREX, the sonic point moves to the corner of the blunt body and affects the pressure distribution throughout the subsonic flow. In the case of the OREX with $\alpha_N = 50^o$, the pressure coefficient shows over expanded flow. The spherical blunted cone-flare configuration ($\theta = 20^o$) gives under expanded flow as seen in figure 36. A sudden drop in C_p is observed on the shoulder of the module followed by a negative C_p variation in the base region. A low pressure is formed immediately downstream of the base which is characterized by a low-speed recirculating flow region which can be attributed to fill-up the growing space between the shock wave and the body. In the region, C_p is decreasing with increasing M_∞.

An oscillating flow over the reentry configuration analyzed using numerical simulation [98]. Flow oscillations over Apollo and OREX are numerically simulated by solving unsteady, axisymmetric, compressible Navier-Stokes equations for $M_\infty = 1.2$. The periodicity of the flow fields in the base region are 8.33 Hz and 1.04 Hz, respectively for the Apollo and the OREX capsules. The cycle of motion of the flow field over the Apollo is 8 times faster as compared the OREX. The geometrical configuration of the reentry can be selected based on the aerodynamic and mission requirements.

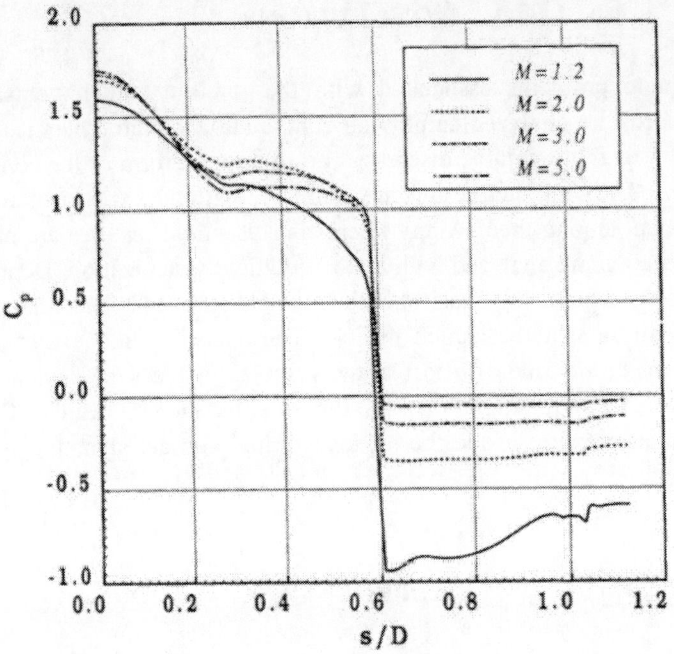

(a) OREX

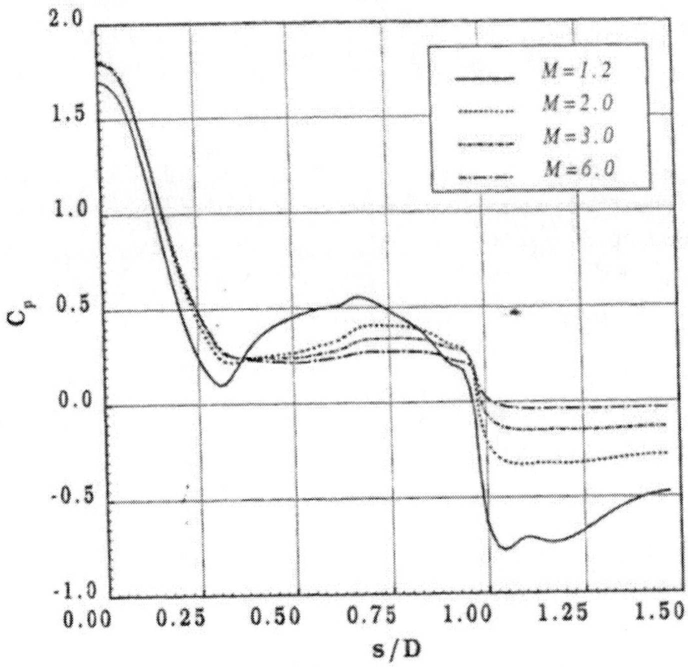

(b) Spherical blunt cone flare

Figure 36. Variation of pressure distribution along the surface of the reentry capsules.

Conclusions

The aerodynamic problems associated with the satellite launch vehicle and reentry capsule at high-speeds are analyzed employing computational fluid dynamics approach. The numerical method uses finite volume discretization of integral form of the compressible fluid dynamics equations. The computed results are compared with the experimental data and show fairly good agreement among them. Many interesting flowfield features are observed during the numerical simulation are analyzed with aid of flowfield visualization. Depressurization of the heat shield is carried out using quasi-one-dimensional approach. The numerical results are compared with flight data of the launch vehicle. The analysis will be useful to obtain the flowfield simulations at the time of liftoff of the vehicle, to select the payload shroud of the heat shield, process the post flight data of the satellite launch vehicle, design limit of maximum differential pressure inside the payload faring, and aerodynamic consideration of the reentry capsule.

References

[1] Lamont, P. J. and Hunt, B. L. The impingement of under expanded axisymmetric jets on wedges, *Journal of Fluid Mechanics*, Vol. 76, Part 2, 1976, pp. 307 – 336.

[2] Kim, K.-H. and Chang, K.-S. Three-dimensional structure of a supersonic jet impinging on an inclined plate, *AIAA Journal*, Vol. 31, No.5, 1994, pp. 778 – 782.

[3] Nakai, Y., Fujimatsu, N., and Fujii, K. Experimental study of under expanded supersonic jet impinging on an inclined plate, *AIAA Journal*, Vol. 44, No. 11, 2006, pp. 2691 – 2699.

[4] Wu, J., Tang, L., Luke, E. A., Tong, X – L. and Cinnella, P. Comprehensive numerical study of jet flow impingement over flat plates, *Journal of Spacecraft and Rockets*, Vol. 39, No. 3, 2002, pp. 357 – 366.

[5] Anon, *Acoustic loads generated by the propulsion system*, NASA Space Vehicle Design Criteria (Structures), NASA SP 8072, June 1971.

[6] Prasad, J. K., Mehta, R. C. and Sreekanth, A. K. Impingement of supersonic jets on an axisymmetric deflector, *AIAA Journal*, Vo.32, No.7. July 1994, pp.1535-1538.

[7] Prasad, J. K., Mehta, R. C. and Sreekanth, A. K. *Impingement of supersonic jets on a wedge and axisymmetric deflector, in the proceedings of 6th Asian Congress of Fluid Mechanics*, Nanyang Technological University, Singapore, pp.983-987, May 22-26, 1995.

[8] Mehta, R. C., Pandya, M. J. and Jayachandran, T. Euler calculations for diffuser flow field, free jets and impingement jets, *International Journal of Numerical Methods of Heat and Fluid Flows*, Vol.4, No.4, May 1995, pp. 287-300.

[9] Mehta, R. C. and Prasad, J. K. Investigation of supersonic free jets emanating from convergent-divergent nozzle, *International Journal of Computational Fluid Dynamics*, Vol. 10, 1998, pp. 61-71.

[10] Park, D.C. A note on Prandtl formula for the wave length of supersonic gas jet, *Quarterly Journal of Mechanics and Applied Mathematics*, Vol. 3, No. 2, 1950, pp. 173 -181.

[11] Love, E. S., Grigsby, C. E., Lee, L. P. and Woodling, M. J. *Experimenatl and theoretical studies of axisymmetric free jets*, NASA TR-6, 1959.

[12] Abdel-Fattah, A. M. Discrete tone emission from high pressure ratio supersonic jets from convergent-divergent nozzles, *AIAA Journal*, Vol. 26, No. 3, 1988, pp. 283 – 291.

[13] Mehta, R. C. and Prasad, J. K. Flow structure of a supersonic jet impinging on an axisymmetric deflector, *Indian Journal of Technology and Materials Sciences*, Vol. 4, Oct. 1997, pp. 178-188.

[14] Mehta, R. C. and Prasad, J. K. Numerical analysis of flow fields of axisymmetric supersonic free jets, *Computational Fluid Dynamics JOURNAL*, Vol. 12, No. 1, April 2003, pp. 115 – 120.

[15] Mehta, R. C. and Prasad, J. K. Estimation of shock-cell structure of axisymmetric free jets, *Indian Journal of Engineering and Material Sciences*, Vol. 3, August 1996, pp. 141 – 147.

[16] Powell, A. *On the mechanism of chocked jet noise*, Proceedings of the Physical Society, 866, 1953, pp. 1039 – 1056.

[17] Mehta, R. C. Numerical simulation of supersonic turbulent jets impinging on an axisymmetric deflector, *Indian Journal of Engineering and Material Sciences*, Vol. 9, August 2002, pp. 227 – 236.

[18] Prasad, J. K., Mehta, R. C. and Sreekanth, A. K. Experimental study of overexpanded supersonic jet impingement on a double wedge deflector, *Journal of the Royal Aeronautical Society*, June/July 1993, pp. 209 – 214.

[19] Hsieh, T. Hemisphere-cylinder in low supersonic flow, *AIAA Journal* Vol. 13, No. 12, Dec. 1973, pp. 1551 – 1552.

[20] Hsieh, T. Hemisphere-cylinder in transonic flow, $M_\infty = 0.7 \sim 1.0$, *AIAA Journal* Vol. 13, No. 10, 1975, pp. 1411 – 1413.

[21] Baldwin, B. S. and Lomax, H. *Thin layer approximation and algebraic model for separated turbulent flow*, AIAA paper 78-257, 1978.

[22] Mehta, R. C. *Numerical investigation of viscous flow over a hemisphere-cylinder*, Acta Mechanica, Vol. 128, 1998, pp. 49-58.

[23] Mehta, R. C. Numerical simulation of wall pressure fluctuations on hemisphere cylinder at transonic Mach numbers, *Computational Fluid Dynamics JOURNAL*, Vol. 8, No. 4, Jan. 2000, pp. 511-520.

[24] Anon. *Buffeting During Atmospheric Ascent*, NASA Special Vehicle Design Criteria (Structures), NASA SP-8001, Nov. 1970.

[25] Mehta, R. C. *Comparative Study of Surface Pressure Fluctuations Over Bulbous Heat Shield at Mach 0.95*, Computers and Fluids, Vol. 30, No. 6, June 2001, pp. 689-709.

[26] Mehta, R. C. Solution of viscous flow over heat shield using structured grid, *Aerospace Science and Technology*, Vol. 2, 1998, pp. 415-423.

[27] Mehta, R. C. Transonic flow simulation for a bulbous heat shield, *Journal of Spacecraft and Rockets*, Vol.34, No.4, July-Aug. 1997, pp.561-564.

[28] Mehta, R. C. Flowfield over bulbous heat shield in transonic and low supersonic speeds, *Journal of Spacecraft and Rockets*, Vol. 35, No. 1, Jan-Feb 1998, pp.102-105.

[29] Mehta, R. C. Numerical studies over a bulbous payload shroud in transonic and low supersonic Mach numbers, *AIAA Paper* 97-2256, 1997.

[30] Coe, C. F. *The effects of some variations in launch vehicle nose shape on steady and fluctuating pressures at transonic speeds*, NASA TM X 645, March 1962.

[31] Rainey, A. G. Progress on the launch vehicle buffeting problem, *Journal of Spacecraft and Rockets*, Vol. 2, No. 3, 1965, pp. 289 – 299.

[32] Dotson, K. W., Baker, R. L. and Bywater, R. L. Launch vehicle self-sustained oscillation from aero-elastic coupling, Part 2: Analysis, *Journal of Spacecraft and Rockets*, Vol. 35, No. 3, 1998, pp. 365 - 373

[33] Mehta, R. C. and Sastri, V. M. K. Numerical analysis of wall pressure and heat flux fluctuations in shock/turbulent boundary layer interaction, *International Journal of Numerical Methods in Fluids*, Vol.16, No.8, April 1993, pp.739-753.

[34] Mehta, R. C. Unsteady pressure fluctuations in the boat tail region of a heat shield, *Aerospace Science and Technology*, Vol. 4, 2000, pp. 91 – 102.

[35] Marple Jr., Digita, S. L., *l Spectral Analysis with applications*, Prentice-Hall inc., NJ, 1987.

[36] Mehta, R. C., Wall pressure fluctuations over a bulbous heat shield of a satellite launch vehicle, *Acta Mechanica*, Vol. 137, 1999, pp. 13 – 31.

[37] Friberg, E. G. and Walchner, O. Pressure correlation for blunted slender cones, *AIAA Journal*, Vol. 7, No. 8, 1969, pp.1618 – 1619.

[38] Owens, R. V. *Aerodynamic characteristics of spherical blunted cones at Mach numbers from 05 to 5.0*, NASA TN D 3088, 1965.

[39] Muraca, R. J. External and internal air loads on sounding rockets, *Journal of Spacecraft and Rockets*, Vol. 4, No. 9, 1967, pp. 1207 – 1210.

[40] Mehta, R. C. and Sastry, M. S. Computation of Aerodynamic load and coefficient for satellite launch vehicle, *Indian Journal of Engineering and Material Sciences*, Vol. 4, April 1997, pp. 41 – 48.

[41] Reding, J. P., Guenther, R., and Richer, B. J. Unsteady aerodynamic considerations in the design of a drag reduction spike, *Journal of Spacecraft and Rockets*, Vol. 14, 1977, pp. 705 – 710.

[42] Zorea, C. and Rom, J. Effects of a spike on the drag and on the aerodynamic stability of blunt bodies in supersonic flow, *Journal of Spacecraft and Rockets*, Vol. 7, 1990, pp. 1017 – 1019.

[43] Mehta, R. C. and Jayachandran, T. Navier Stokes solution for a heat shield with and without a forward facing spike, Computers and Fluids, Vol. 26, No. 7, 1997, pp. 741 – 754.

[44] Mehta, R. C. Heat transfer study of high speed flow over a spiked blunt body, *International Journal of Numerical Methods in Heat and Fluid Flow*, Vol. 10, No. 7, 2000, pp. 750 – 769.

[45] Crawford D. H., *Investigation of flow over a spiked-nose hemisphere-cylinder at a Mach number of 6.8*, NASA TN D 118, 1959.

[46] Kobayashi, H., Maru, Y. and Fukiba, K. Experimental study on aerodynamic characteristics of telescopic aerospikes with multiple disks, *Journal of Spacecraft and Rockets*, Vol. 44, No. 1, 2007, pp. 33 – 41.

[47] Reding, J. P., Guenther, R. A., and Jecmen, D. M. Scale effects on fluctuating pressures in spike-induced flow separation, *Journal of Spacecraft and Rockets*, Vol. 17, No. 2, 1980, pp. 112 – 118.

[48] Motoyama, N., Mihara, K., Miyajima, R., Watanuki, T. and Kubota, *H. Thermal protection and drag reduction with use of spike in hypersonic flow*, AIAA paper 2001-1828, 2001.

[49] Feszty, D., Badcock, K. J. and Richards, B. E. *Numerical simulation of hysteresis phenomenon in high-speed spiked body*, AIAA paper 2000 – 0141, 2000.

[50] Calarese, W. and Hankey, W. L. Modes of shock wave oscillations on spike-tipped bodies, *AIAA Journal*, Vol. 23, No. 2, 1985, pp. 185 – 192.

[51] Shang, J. S., Smith, R. E. and Hankey, W. L. *Flow oscillations of spike-tipped bodies,* AIAA paper 80 – 0062, 1980.

[52] Raleigh, L. *On the stability or instability of certain fluid motion*, Scientific papers, Vol. 1, Cambridge University Press, 1880, pp. 474 -487.

[53] Yamauchi, M., Fujii, K. and Higashino, F. Numerical *investigation of supersonic flows around spiked blunt body, Journal* of Spacecraft and Rockets, Vol. 32, No. 1, 1995, pp. 32 - 42.

[54] Mehta, R. C., Kalimuthu, R. and Rathakrishnan, E. *Numerical simulation of flow field over aerodisc at Mach 6*, 34th National Conference on Fluid Mechanics and Fluid Power, BIT Mesra, Ranchi, India, Dec. 10-12, 2007.

[55] Kreyszig, E. *Advanced Engineering mathematics*, 5th Ed. Wiley, New Delhi, p. 135.

[56] Mehta, R. C. *Numerical analysis of pressure oscillations over axisymmetric spiked blunt bodies at Mach 6.8*, Shock Waves, Vol. 11, No. 6, May 2002, pp. 431-440.

[57] Mehta, R. C. Pressure oscillations over a spiked blunt body at hypersonic Mach number, *Computational Fluid Dynamics Journal*, Vol. 9, No. 2, July 2000, pp. 88-95.

[58] Mehta, R. C. *Limit cycle of oscillating flow over a spiked blunt body, in the proceedings of the 4th Asian Computational Fluid Dynamics Conference*, Mianyang, Sichuan, China, September 19-21, 2000.

[59] Mehta, R. C. Numerical heat transfer study over spiked blunt bodies at Mach 6.8, *Journal of Spacecraft and Rockets*, Vol. 37, No. 5, 2000, pp. 700 – 703.

[60] Mehta, R. C. Peak heating for reattchament of separated flow on a spiked blunt body, *Heat and Mass Transfer*, Vol. 36, 2000, pp. 277 – 283.

[61] Mehta, R. C. Numerical analysis of pressure oscillations over axisymmetric spiked blunt bodies at Mach 6.8, *Shock Waves*, Vol. 11, No. 6, May 2002, pp. 431-440.

[62] Mehta, R. C. Pressure oscillations over a spiked blunt body at hypersonic Mach number, *Computational Fluid Dynamics Journal*, Vol. 9, No. 2, July 2000, pp. 88-95.

[63] Hillije, E. R., and Nelson, R. L. *Ascent air data system results from the space shuttle flight test program – Lessons learned,* compiled by J. P. Arrington, NASA CR 2283, 1983.

[64] Rediniotis, O. K. and Vijaygopal, R. Miniature muti-hole pressure probes and their neural network based calibrations, *AIAA Journal*, Vol. 37, No. 6, June, 1999, pp. 666 – 674.

[65] Mehta, R. C. Flow direction estimation based on computed and flight measured surface pressure, *Computational Fluid Dynamics JOURNAL*, Vol. 12, No. 3, Oct. 2003, pp. 555-561.

[66] *Payload Accommodation in International Reference Guide to Space Launch Systems*, Edited by S. J. Isakowitz, J. B. Hopkins, J. P. Hopkins, Jr., Fourth Edition, AIAA Publication, July 2004.

[67] Anon. *Compartment Venting: Space Vehicle Design Criteria (Structures),* NASA SP-8060, Nov. 1970.

[68] Mironer, A., and Regan, F. *Venting of Space-Shuttle Payloads*, AIAA paper 83-2600 (1983).

[69] Murri, B. W. *Payload Venting in Worst Case Shuttle Environments*, AIAA paper 87-1598 (1987).

[70] Sanz-Andres, A., Satiago-Prowald, J., and Ayuso-Berea, A., Spacecraft Launch Depressurization loads, *Journal of Spacecraft and Rockets* Vol. 34, No.6, 1997, pp.805 - 810.

[71] Huseman, P. G., and Chern, S-Y. *Discharge Coefficient Analysis for Titan IV Payload Fairing Vents*, AIAA paper 97-2068 (1997).

[72] John, A. L., and Jones, M. L. *Venting characteristics of gaseous helium and nitrogen discharging into a freestream at Mach number from 0.60 to 1.57*, NASA TM X-2995 (1974).

[73] Fay, J. F., and Hengel, J. E. *Pressure Dither in Venting Analyses*, AIAA paper 93-3014 (1993).

[74] Lutfi, H. S., and Nieder, R. L. *Space Shuttle Obiter Venting – Lesson Learned, complied by J. P. Arrington*, NASA CR 2283, 1983.

[75] Mehta, R. C. Inverse Venting Problem for Estimation of the Discharge Coefficient in a Satellite Launch Vehicle, *Journal of Aerospace Engineering*, Vol. 217, No. G6, 2003, pp. 277-281.

[76] Mehta, R. C. Quasi-one-dimensional numerical analysis of payload venting of satellite launch vehicle, *Journal of Spacecraft and Rockets*, Vol. 45, No. 2, 2008, pp.412-414.

[77] Brower, T. L. Titan launch vehicle: Ground Test History, *Journal of Spacecraft and Rockets*, Vol. 43, No. 1, 2006, pp. 147 - 160.

[78] Gnoffo, P. A., Price, J. M. and Braum, R. D. *On the computation of near wake aerobrake flow fields*, AIAA paper 91-1371, June 1991.

[79] Grasso, F. and Marinir, M. *Solution of hypersonic flows with total variation diminishing multigrid techniques*, Computer and Fluids, Vol. 23, No. 5, 1995, pp. 571-592.

[80] Venkatapathy, E., Palmer, G., and Prabhu, D. K. *AFE base computations*, AIAA paper 91-1372, June 1991.

[81] Allen, J. S. and Cheng, S. I. Numerical solution of the compressible Navier-Stokes equations for the near wake, *Physics of Fluids*, Vol. 13, No. 1, 1970, pp. 37-52.

[82] Weinbaum, S. Rapid expansion of a supersonic boundary layer and its applications to the near wake, *AIAA Journal*, Vol. 4, No. 2, 1966, pp. 217-226.

[83] Osu, H., Abe, T., Ohnishi, Y. Sasoh, A. and Takayama, K., Numerical investigation of high enthalpy flow generated by expansion tube, *AIAA Journal*, Vol. 40, No. 12, 2002, pp. 2423-2430.

[84] Grasso, F. and Pettinelli, C. Analysis of laminar near wake hypersonic flow, *Journal of Spacecraft and Rockets*, Vol. 32, No. 6, 1995, pp. 970-980.

[85] Tai, C-S. and Kao, A. F. Navier-Stokes solver for hypersonic flow over a slender cone, *Journal of Spacecraft and Rockets*, Vol. 31, No. 1, 1994, pp. 215-222.

[86] Lamb, J. P. and Oberkampf, W. L. Review and development of base pressure and base heating correlations in supersonic flow, *Journal of Spacecraft and Rockets*, Vol. 32, No. 1, 1995, pp. 8-23.

[87] Wood, A. W., Gnoffo, P. A. and Rault, D.F.G. Aerodynamic analysis of commercial experiment transport reentry capsule, *Journal of Spacecraft and Rockets*, Vol. 33, No. 5, 1996, pp. 643-646.

[88] Teramoto, S., Hiraki, K., and Fujii, K. Numerical analysis of dynamic stability of a reentry capsule at transonic speed, *AIAA Journal*, Vol. 39, No. 4, April 2001, pp. 646-653.

[89] Yamamoto, Y. and Yoshioka, M. *CFD and FEM coupling analysis of OREX aero-thermodynamic flight data*, AIAA 95-2087, 1995.

[90] Tam, L. T. *LU-SGS implicit scheme for entry vehicle flow computation and comparison with aerodynamic data*, AIAA 92-2671 CP, 1992.

[91] Liever, P. A., Habchi, S. D., Burnell, S. I. and Lingard, J. S. Computational fluid dynamics prediction of the Beagle 2 aerodynamic data base, *Journal of Spacecraft and Rockets*, Vol. 40, No. 5, 2003, pp. 632-638.

[92] Haas, B. L. and Venkatapathy, E. *Mars path finder computations including base heating predictions*, AIAA 95-2086, 1995.

[93] Mehta, R. C. *Aerodynamic Drag Coefficient for Various Reentry Configurations at High Speed*, AIAA paper 2006-3179, 2006.

[94] Mehta, R. C. Computation of flow field over reentry capsules at supersonic Mach numbers, *Computational Fluid Dynamics JOURNAL*, Vol. 13, No. 3, Oct. 2004, pp. 585-596.

[95] Mehta, R. C. *Numerical simulation of supersonic flow past reentry capsules*, Shock Waves, Vol. 15, No. 1, March 2006, pp. 31-41.

[96] Mehta, R. C. *A quasi-three-dimensional automatic grid generation method, in the proceedings of the 25th National and International Conference on Fluid Dynamics and Fluid Flow*, IIT, Delhi, Dec. 1998, pp. 89 – 98.

[97] Shang, J. S. *Numerical simulation of wing-fuselage interference aerodynamic*, AIAA paper 81-0084, 1981.

[98] Mehta, R. C. Computational investigation of flow oscillation over reentry capsules, *Computational Fluid Dynamics JOURNAL*, Vol. 15, No. 2, July 2006, pp. 247-260.

In: New Developments in Magnetohydrodynamics … ISBN: 978-1-61942-078-6
Editors: Stephen Talton and Patrick F. Theba 2012 Nova Science Publishers, Inc.

Chapter 4

MHD FLOW OF A MICROPOLAR FLUID WITH ASYMMETRIC WALL TEMPERATURES

S. K. Ravi, A. K. Tiwari and R. K. Singh*
Department of Mathematics, Banaras Hindu University,
Varanasi, India

ABSTRACT

This paper presents the effect of MHD of micropolar fluids with asymmetric wall temperatures in a vertical wall. The transformed boundary layer governing equations are solved by the Matlab software. The dynamic of micropolar fluids has been a popular area of research. As the fluid consist of randomly oriented molecules, and as each volume element of the fluid have translated as well as rotation. The analysis of physical problems in these fluids has revealed several interesting phenomena, which are not found in Newtonian fluids. Increasing the Magnetohydrodynamics parameter tends to decrease the velocity profile while the micro rotation increase as the Magnetohydrodynamics increases in vertical walls.

Keywords: Micropolar fluid; Magnetohydrodynamics; Matlab.

NOMENCLATURE

d	- walls width
g	- gravitational acceleration
Gr	- grashof number
N	- dimensionless angular velocity
j	- microinertia density
m	- wall temperature ratio
R	- vertex viscosity parameter

* Email address: santosravi@rediffmail.com

T'	- temperature
u'	- stream wise velocity
u	- dimensionless stream wise velocity
y'	- transverse coordinate
y	- dimensionless transverse coordinate

Greek Symbols

β	- coefficient of thermal expansion
γ	- spin gradient viscosity
T	- dimensionless temperature
κ	- vortex viscosity
μ	- dynamic viscosity
w'	- angular viscosity
ρ	- density

1. Introduction

The research about micropolar fluids has been of great importance because the characteristics of fluid with suspended particles can't be successfully described by the theory of Newtonian fluids. Micropolar fluids can exhibit the microscopic effect due to the local structure and micromotions of the fluid element. C. Fetecau et al. [1] have analyzed the unsteady flow of a second grade fluid between two side walls perpendicular to a plate. As regards the concept and analysis of non-Newtonian fluid flow, several theories have been reported but magnetic effect is very useful in the non-Newtonian fluids. In the present century, with the increase in technological advancements in the area of non-Newtonian fluids (molten plastics, synthetic fibers, polymeric liquids, drilling of petroleum, blood and food stuffs or slurries), akin attempts have been made to solve the extremely complex equations of motion. Various models have been proposed in past to describe the behavior of non-Newtonian fluids.

M. Khan et al. [2] have examined the MHD flows of a second grade fluid between two side walls perpendicular to the plate through a porous medium. R. Cortell [3] studied on the viscous flow and heat transfer over a nonlinearly stretching sheet. R. Nazar et al. [4] have analyzed the stagnation point flow of a micropolar fluid towards a stretching sheet. C. Fetecau et al. [5] studied on the starting solutions for the motion of a second grade fluid due to longitudinal and torsional oscillations of a circular cylinder. C. Fetecau [6] analyzed the unsteady flows of Oldroyd-B fluids in a channel of rectangular cross-section. C. Fetecau et al. [7] studied on the starting solutions for some unsteady unidirectional flows of a second grade fluid. W.C. Tan et al. [8] have studied on Stokes first problem for an Oldroyd-B fluid in a porous half space. S. Baris et al. [9] studied three-dimensional stagnation point flow of a second grade fluid towards a moving plate.

C. Y. Cheng [10] investigated the fully developed natural convection heat and mass transfer of a micropolar fluid in a vertical channel with asymmetric wall temperatures and concentrations. Hayat, T. et al. [11] have considered the influence of Hall current on the rotating oscillating flows of an Oldroyd-B fluid in a porous medium. P.D. Ariel et al. [12] studied on the flow of an elastico-viscous fluid past a stretching sheet with partial slip. Hayat, T. et al. [13] investigated the series solution for the upper-convected Maxwell fluid over a porous stretching plate. R. Cortell [14] investigated the flow and heat transfer of a fluid through a porous medium over a stretching surface with internal heat generation/absorption and suction/blowing. R. Cortell [15] studied on the flow and heat transfer of an electrically conducting fluid of second grade over a stretching sheet subject to suction and to a transverse magnetic field. R. Cortell [16] has investigated the effects of viscous dissipation and work done by deformation on the MHD flow and heat transfer of a viscoelastic fluid over a stretching sheet. R. Cortell [17] has investigated a note on flow and heat transfer of a viscoelastic fluid over a stretching sheet. K. Vajravelu et al. [18] studied on the fluid flow over a nonlinear stretching sheet. E.M. Abo-Eldahab et al. [19] investigated the flow and heat transfer in micropolar fluid past a stretching surface embedded in a non-Darcian porous medium with uniform free stream. Y.Y. Lok et al. [20] have studied on the unsteady boundary layer flow of a micropolar fluid near the forward stagnation point of a plane surface.

A. K. Tiwari [21] has studied on the MHD free convection in a porous medium bounded by a long vertical wavy wall and a parallel flat wall. R. K. Singh et al. [22] have studied on the transient hydromagnetic natural convection between two vertical walls heated/cooled asymmetrically. Anand Kumar et al. [23] have studied on the effect of suction/injection on MHD flat plate thermometer. Chamkha AJ et al. [24] analyzed the thermal radiation effects on MHD forced convection flow adjacent to a non-isothermal wedge in the presence of heat source or sink. S.J. Liao [25, 26] has studied on the solving solitary waves with discontinuity by means of homotopy analysis method and a uniformly valid analytic solution of 2D viscous flow past a semi-infinite flat plate. Y. Tan et al. [27] have studied on the explicit series solution of travelling waves with a front of Fisher equation.S.J. Liao [28, 29] has solved an analytic solution of unsteady boundary-layer flows caused by an impulsively stretching plate, the comparison between the homotopy analysis method and homotopy perturbation method.

S.P. Zhu [30] has solved an exact and explicit solution for the valuation of american put options. Y. Wu et al. [31] have studied on the solving solitary waves with discontinuity by means of the homotopy analysis method. Hayat, T. et al. [32-35] have analyzed the homotopy solution for a generalized second grade fluid past a porous plate, Magnetohydrodynamic flow of an Oldroyd 6-constant fluid, Homotopy analysis of MHD flows of an Oldroyd 8-constant fluid, couette and Poiseuille flows of an Oldroyd 6-constant fluid with magnetic field. Z. Abbas [36] studied on the MHD boundary-layer flow of an upper-convected Maxwell fluid in a porous channel. Sajid, M et al. [37] have analyzed on the analytic solution of the steady flow of a fourth grade fluid. Hayat, T. et al. [38] have studied on the influence of thermal radiation on MHD flow of a second grade fluid. K. R. Rajagopal et al. [39, 40] have analyzed on the flow of viscoelastic fluid over a stretching sheet and an exact solution for the flow of a non-Newtonian fluid past an infinite plate. K. R. Rajagopal et al. [41, 42] have studied on a non-similar boundary layer on a stretching sheet in a non-Newtonian fluid with uniform free stream and on the Oberbeck –Boussinesq approximation.

The present work shows the effect of MHD parameter on the velocity profile and micro-rotation profile. The graphs are plotted by the Matlab software. Increasing the Magnetohydrodynamics parameter tends to decrease the velocity profile while the microrotation increase as the Magnetohydrodynamics increases in vertical channel.

2. Formulation of the Problem

We consider the steady free-convective flow of a viscous and electrically conducting non-Newtonian micropolar fluid between two insulated vertical walls separated by a distance d apart. The inlet temperature is T_0. The temperature of walls at $y' = 0$ and d is T_1' and T_2' respectively. A transverse uniform external magnetic field of strength B_0 is applied perpendicular to the walls. With introducing Boussinesq approximations [42], the governing equation can be obtained as

$$(\mu + \kappa)\frac{d^2u'}{dy'^2} + \kappa\frac{dN'}{dy'} + \rho g\beta\left(T' - T_0'\right) - \frac{\sigma}{\rho}B_0^2u' = 0 \tag{1}$$

$$\gamma\frac{d^2N'}{dy'^2} - \kappa\left(2N' + \frac{du'}{dy'}\right) = 0 \tag{2}$$

$$\frac{d^2T'}{dy'^2} = 0. \tag{3}$$

The boundary conditions for the velocity, angular velocity and temperature for considered model are as follows

$$u' = 0, \quad N' = 0, \; T' = T_1' \text{ at } y' = 0$$

$$u' = 0, \quad N' = 0 \; T' = T_2' \text{ at } y' = d \tag{4}$$

where u', T', N', κ, μ, ρ, β, g are respectively velocity component, temperature, angular velocity, vortex viscosity, dynamic viscosity, fluid density, coefficients for thermal expansion and gravitational acceleration of the micropolar fluid. Additional dimensionless parameter appearing in the above equations is the spin gradient viscosity γ defined by $\gamma = (\mu+0.5\kappa)\, j$ as used [10] where j is also defined in nomenclature.

The dimensionless parameters can be defined and given as

$$Y = \frac{y'}{d}, \; N = \frac{N'd^2\rho}{(G_r\mu)}, \; U = \frac{u'd\rho}{(G_r\mu)}, \; T = \frac{(T' - T_0')}{(T_1' - T_0')},$$

$$G_r = \frac{\beta g(T_1' - T_0')d^3\rho^2}{\mu^2}, \quad m = \frac{(T_2' - T_0')}{(T_1' - T_0')}, \quad M = \frac{\sigma B_0^2 d^2}{\rho\mu} \tag{5}$$

where G_r is the Grashof number

In order to non-dimensional Equation (5), Equations (1), (2) and (3) reduced to

$$(1+R)\frac{d^2U}{dy^2} + R\frac{dN}{dy} + T - MU = 0 \tag{6}$$

$$(1+\frac{R}{2})\frac{d^2N}{dy^2} - DR\,(2N + \frac{du}{dy}) = 0 \tag{7}$$

$$\frac{d^2T}{dy^2} = 0 \tag{8}$$

The boundary conditions in dimensionless form are:

$u = 0, N = 0\ T = 1$ at $y = 0$

$$u = 0\ N = 0\ T = m \text{ at } y = 1 \tag{9}$$

where $m = \frac{(T_2 - T_0')}{(T_1 - T_0')}$ is wall temperature ratio, D and R are material parameters which define as:

$$D = d^2/j \text{ and } R = \kappa/\mu. \tag{10}$$

3. Numerical Solution

The differential equations (6), (7) and (8) with appropriate boundary conditions (9) are solved numerically by using Matlab.

4. Results and Discussion

The system of transformed governing equations (6), (7) and (8) with boundary conditions (9) are solved numerically by using Matlab. Figure 1 plot the velocity profiles for various vortex viscosity parameter (R=0, 0.2, 0.4 and 0.6), m=0, D=0.5 and M=1.0. The magnitude of velocity profile tends to decrease as the vortex viscosity parameter is increased. Sets of respective numerical results are illustrated graphically.

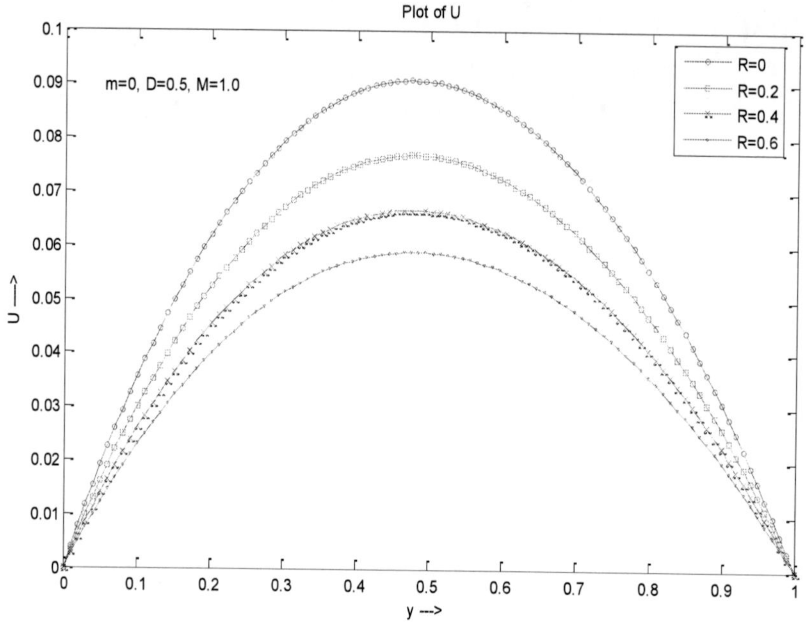

Figure 1. Effects of vortex viscosity parameter on the velocity profile.

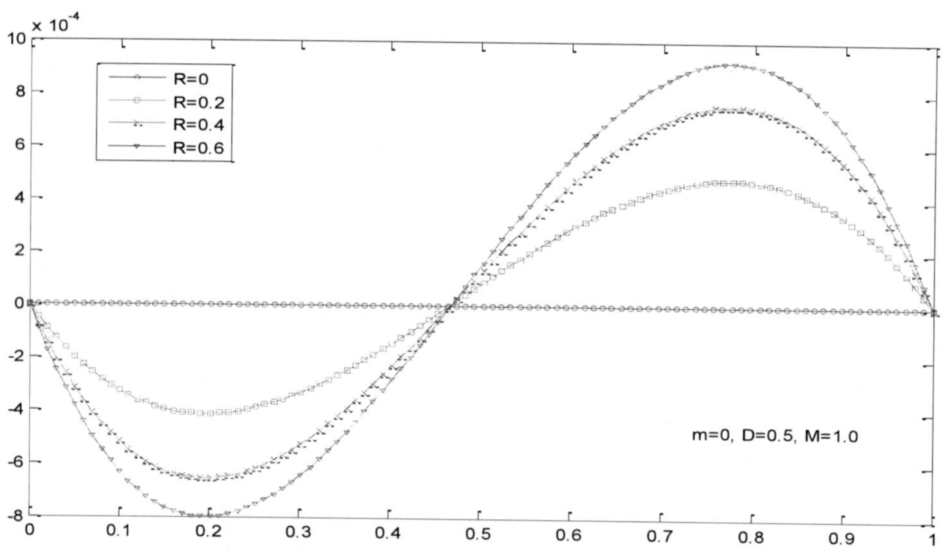

Figure 2. Effects of vortex viscosity parameter on the micro-rotation profile.

Figure 2 shows the microrotation profiles of fluid for R (=0, 0.2, 0.4, 0.6), m=0, D=0.5 and M=1.0. The magnitude of microrotation tends to increase as the vortex viscosity parameter is increased. Microrotion profile is downward direction for first half ($0 \leq y \leq 0.5$) and upward direction second half ($0.5 \leq y \leq 1.0$). Microrotation profile is approximately symmetric about middle region of walls but opposite in nature.

Figure 3 shows the effect of magnetic parameter on the velocity profile. On the various values of magnetic parameter (M=0, 1.0, 2.0, 3.0 and 4.0), m=0, D=0.5 and R=0.1. The maximum values of velocity profile decreases on the increasing of magnetic parameter. Figure 4 Illustrate the microrotation profile of fluid for the case m=0, D=2.0, R=0.2. The magnitude of microrotation decreases as the increasing of magnetic parameter.

Figure 5 indicates the effect of magnetic parameter on the velocity profile. On the various values of magnetic parameter (M= 1.0, 2.0, 3.0 and 4.0), m=0, D=0.5 and R=0. The velocity decreases while the increasing of magnetic parameter.

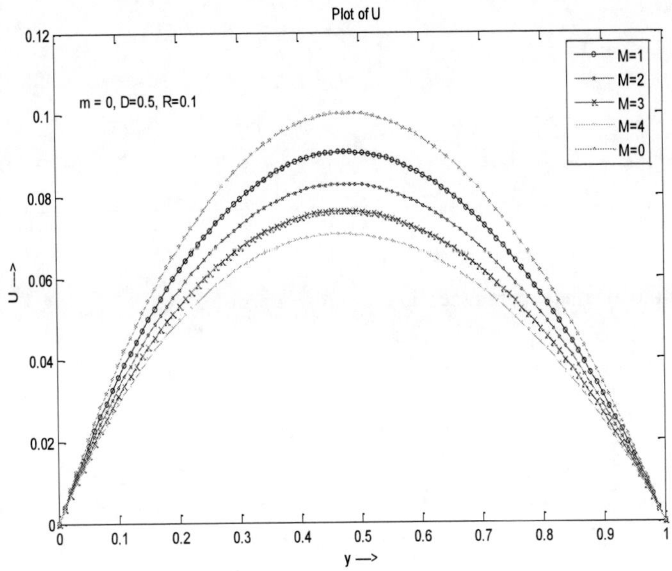

Figure 3. Effects of the magnetic parameter on the velocity profile.

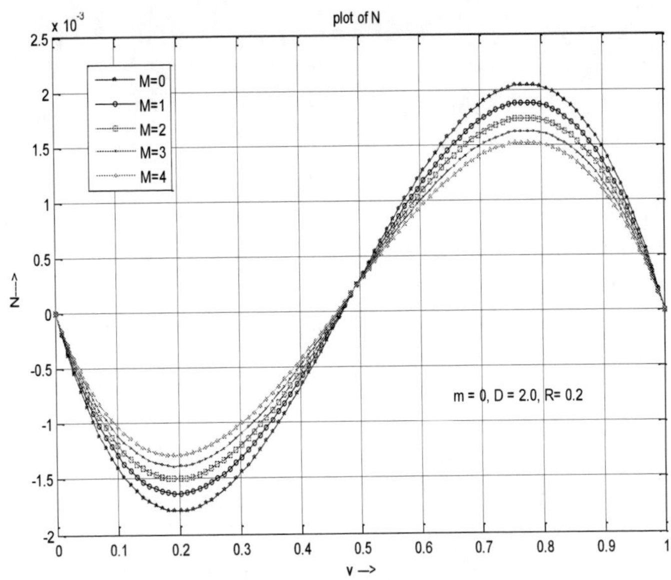

Figure 4. Effect of the magnetic parameter on the micro-rotation profile.

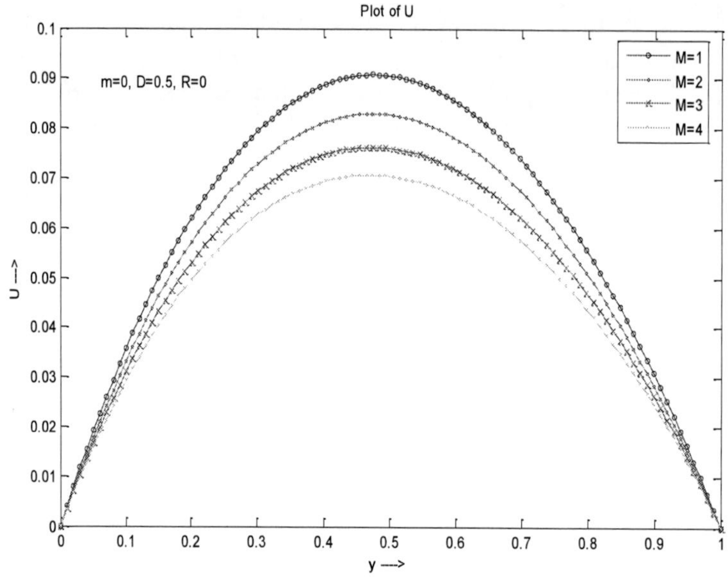

Figure 5. Effect of the magnetic parameter on the velocity profile.

CONCLUSIONS

MHD flow of a micropolar fluid with asymmetric walls temperatures in a vertical channel has been studied. The numerical solutions of many characteristics of micropolar fluids and magnetic parameter have been obtained and studied. Increasing the magnetic parameter tends to decrease the velocity profile as well as magnitude of microrotation.

REFERENCES

[1] Fetecau, C., Hayat, T., Corina Fetecau and Ali, N. *Unsteady flow of a second grade fluid between two side walls perpendicular to a plate*, Nonlinear Anal.RWA 9 (2008) 1236–1252.

[2] Khan, M., Hyder Ali, S., Hayat, T. and Fetecau, C. MHD flows of a second grade fluid between two side walls perpendicular to the plate through a porous medium, *Int. J. Nonlinear Mech.* 43 (4) (2008) 302–319.

[3] Cortell, R. Viscous flow and heat transfer over a nonlinearly stretching sheet, *Appl. Math. Comput.* 184 (2007) 864–873.

[4] Nazar, R., Amin, N., Filip, D. and Pop, I. Stagnation point flow of a micropolar fluid towards a stretching sheet, *International J. Non-Linear Mech.* 39(2004) 1227–1235.

[5] Fetecau, C. and Fetecau, C. Starting solutions for the motion of a second grade fluid due to longitudinal and torsional oscillations of a circular cylinder, *Int. J. Eng. Sci.* 44 (2006) 788–796.

[6] Fetecau, C. and Fetecau, C. Unsteady flows of Oldroyd-B fluids in a channel of rectangular cross-section, *Internat. J. Non-Linear Mech.* 40 (2005)1214–1219.

[7] Fetecau, C. and Fetecau, C. Starting solutions for some unsteady unidirectional flows of a second grade fluid, *Int. J. Eng. Sci.* 43 (2005) 781–789.

[8] Tan, W. C. and Masuoka, T. Stokes first problem for an Oldroyd-B fluid in a porous half space, *Phys. Fluid* 17 (2005) 023101-7.

[9] Baris, S. and Dokuz, M. S. Three-dimensional stagnation point flow of a second grade fluid towards a moving plate, *Int. J. Eng. Sci.* 44 (2006)49–58.

[10] Cheng, C. Y. Fully developed natural convection heat and mass transfer of a micropolar fluid in a vertical channel with asymmetric wall temperatures and concentrations, *International Communications in Heat and Mass Transfer* 33 (2006) 627–635.

[11] Hayat, T., Khan, S. B. and Khan, M. The influence of Hall current on the rotating oscillating flows of an Oldroyd-B fluid in a porous medium, *Non-Linear Dynam.* 47 (2007) 353–362.

[12] Ariel, P. D., Hayat, T. and Asghar, S. The flow of an elastico-viscous fluid past a stretching sheet with partial slip, *Acta Mech.* 187 (2006) 29–35.

[13] Hayat, T., Abbas, Z. and Sajid, M. Series solution for the upper-convected Maxwell fluid over a porous stretching plate, *Phys. Lett. A*. 358 (2006)396–403.

[14] Cortell, R. *Flow and heat transfer of a fluid through a porous medium over a stretching surface with internal heat generation/absorption and suction/blowing, Fluid Dyn.*

[15] Cortell, R. Flow and heat transfer of an electrically conducting fluid of second grade over a stretching sheet subject to suction and to a transverse magnetic field, *Int. J. Heat Mass Transfer 49* (2006) 1851–1856.

[16] Cortell, R. Effects of viscous dissipation and work done by deformation on the MHD flow and heat transfer of a viscoelastic fluid over a stretching sheet, *Phys. Lett. A* 357 (2006) 298–305.

[17] Cortell, R. A note on flow and heat transfer of a viscoelastic fluid over a stretching sheet, *Internat. J. Non-Linear Mech.* 41 (2006) 78–85.

[18] Vajravelu, K. and Cannon, J. R. Fluid flow over a nonlinear stretching sheet, *Appl. Math. Comput.* 181 (2006) 609–618.

[19] Abo-Eldahab, E. M., El and Aziz, M. A. Flow and heat transfer in micropolar fluid past a stretching surface embedded in a non-Darcian porous medium with uniform free stream, *Appl. Math. Comput.* 162 (2005) 881–899.

[20] Lok, Y. Y. Phang, Amin, P. N. and Pop, I. Unsteady boundary layer flow of a micropolar fluid near the forward stagnation point of a plane surface, *Int. J. Eng. Sci. 41* (2003) 173–186.

[21] Tiwari, A. K. MHD Free Convection in a Porous Medium Bounded by a Long Vertical Wavy Wall and a Parallel Flat Wall. *Journal of Magnetohydrodynamics Plasma*, Space Research, 2009, vol. 14(3), ***.

[22] Singh, R. K. and Singh, A. K. Transient hydromagnetic natural convection between two vertical walls heated/cooled asymmetrically. *Journal of Magnetohydrodynamics Plasma*, Space Research, 2009, vol. 14(3), ***.

[23] Anand Kumar and Singh, A. K. Effect of Suction/Injection on MHD Flat Plate Thermometer. *Journal of Magnetohydrodynamics Plasma*, Space Research, 2009, vol. 14(3), ***.

[24] Chamkha, A. J., Mujtaba, M, Quadri, A, and Issa, C. Thermal radiation effects on MHD forced convection flow adjacent to a non-isothermal wedge in the presence of heat source or sink. *Heat Mass Transfer* 2003 (39)305–12.

[25] Wu, W. and Liao, S. J. Solving solitary waves with discontinuity by means of homotopy analysis method, *Chaos Solitons Fractals* 26 (2005)177–185.

[26] Liao, S. J. A uniformly valid analytic solution of 2D viscous flow past a semi-infinite flat plate, *J. Fluid Mech.* 385 (1999) 101–128.

[27] Tan, Y., Xu, H. and Liao, S. J. Explicit series solution of travelling waves with a front of Fisher equation, *Chaos Solitons Fractals* 31 (2007)462–472.

[28] Liao, S. J. *An analytic solution of unsteady boundary-layer flows caused by an impulsively stretching plate*, Comm. Non-Linear Sci. Numer. Simm. 11 (2006) 326–339.

[29] Liao, S. J. Comparison between the homotopy analysis method and homotopy perturbation method, *Appl. Math. Compu*t. 169 (2005)1186–1194.

[30] Zhu, S.P. An exact and explicit solution for the valuation of american put options, *Qunatitative Finance* 6 (2006) 229–242.

[31] Wu, Y., Wang, C. and Liao, S. J. Solving solitary waves with discontinuity by means of the homotopy analysis method, *Choas Solitons Fractals* 23 (2005) 1733–1740.

[32] Hayat, T., and Khan, M. Homotopy solution for a generalized second grade fluid past a porous plate, *Non-Linear Dynam.* 42 (2005) 395–405.

[33] Hayat, T., Khan, M. and Asghar, S. Magnetohydrodynamic flow of an Oldroyd 6-constant fluid, *Appl. Math. Comput.* 155 (2004) 417–425.

[34] Hayat, T., Khan, M. and Asghar, S. Homotopy analysis of MHD flows of an Oldroyd 8-constant fluid, *Acta Mech.* 168 (2004) 213–232.

[35] Hayat, T., Khan, M. and Ayub, M. Couette and Poiseuille flows of an Oldroyd 6-constant fluid with magnetic field, *J. Math. Anal. Appl. 298* (2004)225–244.

[36] Abbas, Z., Sajid, M. and Hayat, T., *MHD boundary-layer flow of an upper-convected* Maxwell fluid in a porous channel, Theor. Comput. Fluid Dyn. 20 (2006) 229–238.

[37] Sajid, M, Hayat, T. and Asghar, S. On the analytic solution of the steady flow of a fourth grade fluid, *Phys. Lett. A*. 355 (2006) 18–26.

[38] Hayat, T., Abbas, Z., Sajid, M and Asghar, S. The influence of thermal radiation on MHD flow of a second grade fluid, *Int. J. Heat Mass Transfer 50* (2007) 931–941.

[39] Rajagopal, K. R., Na, T. Y. and Gupta, A. S. Flow of viscoelastic fluid over a stretching sheet, *Rheol Acta* 23(1984) 213-215.

[40] Rajagopal, K. R. and Gupta, A. S. An exact solution for the flow of a non-Newtonian fluid past an infinite plate, *Meccanica*. 19 (1984) 158-160.

[41] Rajagopal, K. R., Na, T. Y. and Gupta, A. S. A non-similar boundary layer on a stretching sheet in a non-Newtonian fluid with uniform free stream, *J. Math. Phys. Sci.* 21 (1987) 189-200.

[42] Rajagopal, K. R., Ruzicka, M. and Srinivas, A. R. On the Oberbeck –Boussinesq approximation, *Math. Models Methods Appl.Sci.* 6(8) (1996) 1157-1167.

In: New Developments in Magnetohydrodynamics … ISBN: 978-1-61942-078-6
Editors: Stephen Talton and Patrick F. Theba 2012 Nova Science Publishers, Inc.

Chapter 5

NA AND HANSEN'S PROCEDURE FOR SOLVING THE MHD FLOW AROUND A CYLINDER IN CROSSED ELECTRIC AND MAGNETIC FIELDS

Mina B. Abd- el - Malek [*] ***and Samy M. A. El-Mansi***
Department of Mathematics, School of Engineering and Sciences.
The American University in Cairo, Cairo, Egypt

ABSTRACT

An analysis of the magneto hydrodynamic (MHD) flow of a liquid metal around a cylinder of arbitrary electrical conductivity, under the influence of a magnetic field perpendicular to an electric current has been presented. Na and Hansen's procedure as a procedure for reducing the number of independent variables is applied. Finally, the obtained ordinary differential equations with the appropriate corresponding conditions are solved analytically. The flow field, the induced magnetic field, and the resulting forces on the cylinder are calculated.

Keywords: Laminar longitudinal MHD-flow, weak magnetic field, strong magnetic field, similarity analysis, characteristic function, frictional force, electromagnetic force.

1. INTRODUCTION

A synthesis of two classical sciences is represented by magneto- fluid- dynamics: fluid mechanics and electromagnetic field theory. In order to study the motion of compressible fluid, one should take into account their thermodynamic state and thus the object of this new science turns out to be the study of the phenomena related to fluid motion in all their complexity.

[*] Corresponding author: On leave from Dept. Eng. Math. & Phys., Faculty of Engineering, Alexandria University, Alexandria 21544, Egypt Tel.: +20-10-006-6496; Email: minab@aucegypt.edu

The Thames River flows in both directions due to an effect of the action of tides and thus has salt waters which are electrically conducting. Historically, the first investigation in this field dates back to 1836 when Faraday attempted to detect the electric field induced by the motion of the conducting waters in the Earth's magnetic field by connecting two electrodes situated near the two banks of the river. But due to the very low conductivity of the fluid, the results were hardly discernible and Faraday gave up the idea.

A second important stage in the development of this science is represented by the experiments of Hartmann and Lazarus in 1937 who pointed out the magnetic field influence on the motion of a fluid.

In the presence of a strong magnetic field, Mercury flowing through a tube was used as a conducting fluid. Although the experimental results confirmed the theoretical studies, no practical reason for continuing these investigations existed at the time [10].

Astrophysics has posed the first genuine magneto fluid dynamic problems. So in 1940 the Swedish scientist H.Alfvĕn proved that new waves, unknown to both fluid mechanics and electromagnetism, can be propagated through a conducting fluid in a magnetic field.

Following World War II, there was much classified research on nuclear fusion. When in the late 1950's the veil of secrecy was lifted, there was a flood of papers on magneto hydrodynamics (MHD). The important was the discovery of the plasma which replaced liquids as operating media [23].

The purpose of the present paper is to study the physical quantities of the laminar longitudinal MHD-flow which arises when a cylinder of infinite length and arbitrary electrical conductivity is placed into a conducting fluid which is submitted to crossed electric and magnetic fields that are homogeneous at infinity and perpendicular to the axis of the cylinder. Such a problem belongs to the large family of electrically driven flows, which are subjects of theoretical and experimental investigations [14, 18, 20, and 21].

Due to possible metallurgical applications, these flows take a great interest. One of these is electromagnetic separation, a technique where small particles are removed from molten metals by means of electromagnetic forces. The forces on cylinders and spheres of different conductivity in perpendicular electric and magnetic fields are measured by means of model experiments which are performed by Marty and Alemany [20, 21]. They pointed out that the forces depend strongly on the shape of the obstacles and, especially, on the direction of the cylinder axis with respect to the force field. For the theory, three aspects are of great importance: the structure of the flow field; the distribution of the electric current density and the force acting on the body.

In previous theoretical studies, the following particular problems have been considered: the force on a cylinder for asymptotic values of the Hartmann number [20, 21]; the velocity and electric field for an insulating cylinder [26] and the flow field and force for an insulating sphere for small Hartmann number and Stokes approximation [19].

In 1990, Gerbeth et al. [16] offered an analytical and numerical study of the problem. Also, they investigated the physical quantities for arbitrary values of the parameters.

Recently, Abd-el-Malek and El –Mansi [4] applied the technique of Lie group to solve the problem of MHD flow around a cylinder in case of weak magnetic field crossed with electric field.

Lie in 1891 evolved a technique to reduce the number of independent variables in a system of differential equations to obtain invariant solutions by using the infinitesimal groups. This technique is sometimes called the method of group-invariant solutions. For a general discussion of this method, one may refer to[6,25].

Although Lie has obtained numerous results for partial differential equations, only the concept of a similarity transformation, introduced as a systematic method about 50 years after Lie's death by Birkhoff [5], caused broader applications and led to a better understanding of various apparently unrelated results, see [29].

In 1969, Bluman and Cole [7] proposed a generalization to Lie's method. They introduced the so-called non-classical method of group-invariant solutions; see Clarkson and Kruskal [9].

Also; in 1971, Na and Hansen [22] presented a systematic procedure for solving partial differential equations using the infinitesimal group of transformations. For application of Na and Hansen's procedure on diffusion equation, one may refer to [30].

The deductive group methods of Na and Hansen [22] and that of Bluman and Cole [7] are both based upon the invocation of invariance of the partial differential equations under an infinitesimal group of transformations. While Bluman and Cole [7] determine the infinitesimals directly by solving the determining equations, Na and Hansen [22] express the infinitesimals of the group in terms of a single function called the *characteristic function.* The determining equations form a set of linear differential equations, the solution of which gives the transformation function or the infinitesimals of the dependent and independent variables.

Throughout the history of similarity analysis, a variety of problems in science and engineering has been solved. Among these we find the analysis of temperature of thermally stagnant lakes by Boutros et al. [8] in 1999, the steady and unsteady laminar boundary layer flow of a non isothermal vertical circular cylinder by Abd-el-Malek and Badran [1,2] in 1990 and 1991. Many physical applications are illustrated in [3, 4, 9, 27, 28].

In this paper, the similarity solutions of the two - dimensional magneto hydrodynamic (MHD) equations in the case of magnetic field perpendicular to an electric current are obtained. The similarity method of the analysis is Na and Hansen's procedure which is well described in the literature [22, 30].The similarity solutions are obtained by solving ordinary differential equations with only one independent (similarity) variable. The flow field, the induced magnetic field and the resulting forces on the cylinder are calculated.

2. Mathematical Formulation of the Problem

Consider a cylinder with infinite length, of radius "a" and electrical conductivity σ_c, is placed into an infinite incompressible conducting fluid (conductivity σ_f), see Figure 1.

Also, consider the steady flow driven by the interaction of a uniform transverse magnetic field $B_0=B_0e_x$ and a homogeneous electric current $J_0=J_0e_y$. The presence of the obstacle destroys the homogeneity of the current.

Therefore the Lorentz force $f = J\times B$ which acts on every volume is not constant. In general, f consists of two parts:

The irrotational part f_1, i.e. (rot $(f_1) = 0$) can be balanced by a pressure distribution, whereas the second part f_2, i.e. (rot $(f_2) \neq 0$) brings the fluid into motion. Consequently, a pressure force F_P and a frictional force F_f act on the cylinder. Furthermore, the interaction between the current within the cylinder and the magnetic field produces an electromagnetic volume force F_{em} unless the cylinder is an insulator. In order to calculate these forces, it is necessary to solve the corresponding MHD- problem. The magneto-hydrodynamic equations governing the steady flow are, see [16]:

$$\mu\Delta\mathbf{V}+\mathbf{J}\times\mathbf{B}=\mathrm{grad}(p)+\rho(\mathbf{V}\mathrm{grad}(\mathbf{V})), \quad (2.1)$$

$$\Delta\mathbf{B}+\mu_0\,\sigma\,\mathrm{rot}(\mathbf{V}\times\mathbf{B})=0, \quad (2.2)$$

$$\mathrm{div}(\mathbf{V})=0, \quad \mathrm{div}(\mathbf{B})=0\ , \quad (2.3)$$

$$\mu_0\mathbf{J}=\mathrm{rot}(\mathbf{B}), \quad (2.4)$$

where μ, ρ, σ, μ_0 are the dynamic viscosity, density, electrical conductivity and magnetic permeability, respectively, and Δ is the two dimensional Laplacian $\Delta\equiv\partial_x^2+\partial_y^2$.

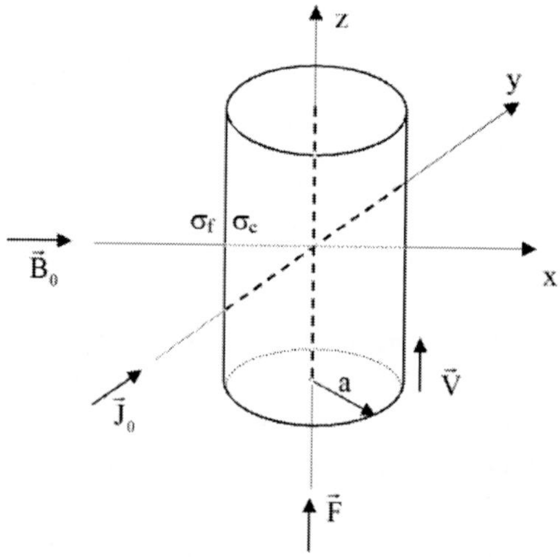

Figure 1. Geometry of the model.

The homogeneous part of the Lorentz force $\mathbf{f}_0= -J_0B_0\mathbf{e_z}$ is balanced by a pressure

$$P_0=-J_0B_0z. \quad (2.5)$$

More precisely, the curl of (2.1) leads, in the present two-dimensional model of an infinite cylinder, to: $\frac{\partial P}{\partial z}$=constant. This constant value is then determined by the boundary conditions at infinity.

Assuming that all other field variables do not depend on z, it is clear that

$$\mathbf{V}=V(x,y)\mathbf{e_z}, \quad (2.6)$$

$$\mathbf{B} = B_0\mathbf{e_x} + B_1(x,y)\mathbf{e_z}, \tag{2.7}$$

$$J = J_0 e_y + J_1, \qquad (J_1.e_z = 0). \tag{2.8}$$

The magnetic field perturbation B_1 inside and outside the cylinder is denoted by B_c and B_f, respectively. The boundary conditions at the cylinder surface are well known as, See [31]:

$$\mathbf{V} = \mathbf{0}, \qquad \mathbf{B_f} = \mathbf{B}_c, \qquad \partial_{\mathbf{n}}\mathbf{B}_f = \partial_n \mathbf{B}_c / \sigma_c, \tag{2.9}$$

where ∂_n denotes the derivative normal to the cylinder surface. They reflect the no slip condition, the continuity of the magnetic field and of the tangential component of the electric field.

At infinity, we require

$$\mathbf{V} \to \mathbf{0}, \qquad \mathbf{J} \to \mathbf{J}_0\mathbf{e_y}, \qquad \mathbf{B}_f \to -\mu_0 J_0 x\mathbf{e_z}. \tag{2.10}$$

This suggests the introduction of the perturbation field B'_f where

$$\mathbf{B}'_f = \mathbf{B}_f + \mu_0 J_0 x\mathbf{e_z}, \tag{2.11}$$

which vanishes at infinity.

The physical quantities become dimensionless by introducing

$$\left.\begin{aligned} X &= x/x^*, \quad Y = y/y^*, \quad Z = z/z^*, \\ v &= V/v^*, \quad b_f = B'_f/b^*, \quad b_c = B_c/b^*, \end{aligned}\right\} \tag{2.12}$$

where x*, y*, z*, v* and b* are reference values which have to be specified.

Incorporating (2.7)-(2.12) into the basic equations (2.1)-(2.4), the following set of equations describing our problem is obtained, see Gerbeth et al. [16]:

$$\Delta\, b_c = 0\ ; \qquad 0 \le r \le 1, \tag{2.13}$$

$$\Delta b_f + M\frac{\partial v}{\partial X} = 0\ ;\ 1 < r < \infty, \tag{2.14}$$

$$\Delta v + M\frac{\partial b_f}{\partial X} = 0\ ;\ 1 < r < \infty, \tag{2.15}$$

$$v=0\ ,\quad b_f=b_c+\frac{1}{M}\cos\phi\ ,\quad \frac{\partial b_f}{\partial r}=\frac{1}{\lambda}\frac{\partial b_c}{\partial r}+\frac{1}{M}\cos\phi\ ;\quad \text{at } r=1, \tag{2.16}$$

$$v\to 0,\quad b_f\to 0,\quad (r\to\infty) \tag{2.17}$$

where r and ϕ denote the polar coordinates $X = r\cos\phi$, $Y = r\sin\phi$, M and λ are the Hartmann number and the dimensionless conductivity of the cylinder defined by:

$$M=B_0\,a\sqrt{\sigma_f/\mu}\quad ,\quad \lambda=\sigma_c/\sigma_f\ . \tag{2.18}$$

The choice

$$x^*=y^*=z^*=a\quad ,\quad v^*=J_0B_0a^2/\mu\quad ,\quad b^*=\mu_0J_0aM\ ,$$

for the reference values ensures that the system (2.13)-(2.17) contains only the two independent parameters M and λ which therefore determine the character of the solutions uniquely.

The system of equations (2.13)-(2.15) is transformed to the following system of equations, using the polar coordinates:

$$r^2\frac{\partial^2 b_c}{\partial r^2}+r\frac{\partial b_c}{\partial r}+\frac{\partial^2 b_c}{\partial\phi^2}=0\quad ;\quad 0\le r\le 1\ , \tag{2.19}$$

$$r^2\frac{\partial^2 b_f}{\partial r^2}+r\frac{\partial b_f}{\partial r}+\frac{\partial^2 b_f}{\partial\phi^2}+Mr^2\cos\phi\frac{\partial v}{\partial r}-Mr\sin\phi\frac{\partial v}{\partial\phi}=0\quad ;\quad 1<r<\infty\ , \tag{2.20}$$

$$r^2\frac{\partial^2 v}{\partial r^2}+r\frac{\partial v}{\partial r}+\frac{\partial^2 v}{\partial\phi^2}+Mr^2\cos\phi\frac{\partial b_f}{\partial r}-Mr\sin\phi\frac{\partial b_f}{\partial\phi}=0\quad ;\quad 1<r<\infty\ , \tag{2.21}$$

with the following boundary conditions:

$$v=0\ ,\quad b_f=b_c+\frac{1}{M}\cos\phi\ ,\quad \frac{\partial b_f}{\partial r}=\frac{1}{\lambda}\frac{\partial b_c}{\partial r}+\frac{1}{M}\cos\phi\ ;\quad r=1, \tag{2.22}$$

$$v\to 0\ ,\quad b_f\to 0\quad \text{as } r\to\infty\ . \tag{2.23}$$

3. Similarity Solution of the Magnetic Field Perturbation "B_C" Equation

3.1. The Group Systematic Formulation

Let us introduce the notations

$$p_1 = \frac{\partial b_c}{\partial \phi},\quad p_2 = \frac{\partial b_c}{\partial r},\quad p_{11} = \frac{\partial^2 b_c}{\partial \phi^2},\quad p_{22} = \frac{\partial^2 b_c}{\partial r^2} \text{ and } p_{12} = \frac{\partial^2 b_c}{\partial r \partial \phi}. \tag{3.1}$$

Hence, the infinitesimal group of transformation can be expressed in terms of these notations as follows:

$$\bar{r} = r + \varepsilon \zeta^{r}(r,\phi,b_c) + o(\varepsilon^2) , \tag{3.2}$$

$$\bar{\phi} = \phi + \varepsilon \zeta^{\phi}(r,\phi,b_c) + o(\varepsilon^2) , \tag{3.3}$$

$$\bar{b}_c = b_c + \varepsilon \eta^{b_c}(r,\phi,b_c) + o(\varepsilon^2) , \tag{3.4}$$

$$\overline{p_1} = p_1 + \varepsilon \pi_1(r,\phi,b_c,p_1,p_2) + 0(\varepsilon^2), \tag{3.5}$$

$$\overline{p_2} = p_2 + \varepsilon \pi_2(r,\phi,b_c,p_1,p_2) + 0(\varepsilon^2), \tag{3.6}$$

$$\overline{p_{11}} = p_{11} + \varepsilon \pi_{11}(r,\phi,b_c,p_1,p_2,p_{11},p_{12},p_{22}) + 0(\varepsilon^2), \tag{3.7}$$

$$\overline{p_{12}} = p_{12} + \varepsilon \pi_{12}(r,\phi,b_c,p_1,p_2,p_{11},p_{12},p_{22}) + 0(\varepsilon^2), \tag{3.8}$$

$$\overline{p_{22}} = p_{22} + \varepsilon \pi_{22}(r,\phi,b_c,p_1,p_2,p_{11},p_{12},p_{22}) + 0(\varepsilon^2), \tag{3.9}$$

where $\zeta^r, \zeta^\phi, \eta^{b_c}, \pi_1, \pi_2, \pi_{11}, \pi_{12}$ and π_{22} are the infinitesimals of the group of transformations defined as follows in terms of the characteristic function W:

$$W = \zeta^{\phi} P_1 + \zeta^{r} P_2 - \eta^{b_c}, \tag{3.10}$$

$$\zeta^{\phi} = \frac{\partial W}{\partial p_1}, \tag{3.11}$$

$$\zeta^{r} = \frac{\partial W}{\partial p_2}, \tag{3.12}$$

$$\eta^{b_c} = p_1 \frac{\partial W}{\partial p_1} + p_2 \frac{\partial W}{\partial p_2} - W, \tag{3.13}$$

$$-\pi_1 = \frac{\partial W}{\partial \phi} + p_1 \frac{\partial W}{\partial b_c}, \tag{3.14}$$

$$-\pi_2 = \frac{\partial W}{\partial r} + p_2 \frac{\partial W}{\partial b_c}, \tag{3.15}$$

$$-\pi_{11} = \frac{\partial^2 W}{\partial \phi^2} + 2p_1 \frac{\partial^2 W}{\partial \phi \partial b_c} + p_1^2 \frac{\partial^2 W}{\partial b_c^2} + 2p_{12}\left(\frac{\partial^2 W}{\partial \phi \partial p_2} + \frac{\partial^2 W}{\partial b_c \partial p_2}\right) + 2p_2 \frac{\partial^2 W}{\partial \phi \partial p_1} +$$

$$p_1 p_2 \frac{\partial^2 W}{\partial b_c \partial p_1} + p_{12}^2 \frac{\partial^2 W}{\partial p_2^2} + 2p_{12}p_2 \frac{\partial^2 W}{\partial p_1 \partial p_2} + p_2^2 \frac{\partial^2 W}{\partial p_1^2} + p_2 \frac{\partial W}{\partial b_c}, \tag{3.16}$$

$$-\pi_{22} = \frac{\partial^2 W}{\partial r^2} + 2p_2 \frac{\partial^2 W}{\partial r \partial b_c} + p_2^2 \frac{\partial^2 W}{\partial b_c^2} + 2p_{12}\left(\frac{\partial^2 W}{\partial r \partial p_1} + \frac{\partial^2 W}{\partial b_c \partial p_1}\right) + 2p_1 \frac{\partial^2 W}{\partial r \partial p_2} +$$

$$p_1 p_2 \frac{\partial^2 W}{\partial b_c \partial p_2} + p_{12}^2 \frac{\partial^2 W}{\partial p_1^2} + 2p_{12}p_1 \frac{\partial^2 W}{\partial p_1 \partial p_2} + p_1^2 \frac{\partial^2 W}{\partial p_1^2} + p_1 \frac{\partial W}{\partial b_c}, \tag{3.17}$$

where $W = W(r, \phi, b_c, p_1, p_2)$.

3.2. Invariance Analysis

Under the infinitesimal group of transformations, a partial differential equation f=0 will be invariant if

$$Uf = 0, \tag{3.18}$$

or, in expanded form, we have

$$\zeta^{\phi}\frac{\partial f}{\partial \phi}+\zeta^{r}\frac{\partial f}{\partial r}+\eta^{b_c}\frac{\partial f}{\partial b_c}+\pi_1\frac{\partial f}{\partial p_1}+\pi_2\frac{\partial f}{\partial p_2}+\pi_{11}\frac{\partial f}{\partial p_{11}}+\pi_{12}\frac{\partial f}{\partial p_{12}}+\pi_{22}\frac{\partial f}{\partial p_{22}}=0 \quad . (3.19)$$

Equation (2.19) can be written in terms of the notations (3.1) as follows:

$$f = r^2 p_{22} + r p_2 + p_{11} . \quad (3.20)$$

from which, and using (3.19) we have

$$(2rp_{22}+p_2)\frac{\partial W}{\partial p_2}+r^2\pi_{22}+r\pi_2+\pi_{11}=0 \quad . \quad (3.21)$$

Substituting for π_2, π_{11} and π_{22}, we get

$$(2rp_{22}+p_2)\frac{\partial W}{\partial p_2}-r\left(\frac{\partial W}{\partial r}+p_2\frac{\partial W}{\partial b_c}\right)-$$

$$r^2\left[\frac{\partial^2 W}{\partial r^2}+2p_2\frac{\partial^2 W}{\partial r\partial b_c}+p_2^2\frac{\partial^2 W}{\partial b_c^2}+2p_{12}\left(\frac{\partial^2 W}{\partial r\partial p_1}+\frac{\partial^2 W}{\partial b_c\partial p_1}\right)+2p_1\frac{\partial^2 W}{\partial r\partial p_2}\right.$$

$$\left.+p_1p_2\frac{\partial^2 W}{\partial b_c\partial p_2}+p_{12}^2\frac{\partial^2 W}{\partial p_1^2}+2p_{12}p_1\frac{\partial^2 W}{\partial p_1\partial p_2}+p_1^2\frac{\partial^2 W}{\partial p_1^2}+p_1\frac{\partial W}{\partial b_c}\right]-$$

$$\left[\frac{\partial^2 W}{\partial \phi^2}+2p_1\frac{\partial^2 W}{\partial \phi\partial b_c}+p_1^2\frac{\partial^2 W}{\partial b_c^2}+2p_{12}\left(\frac{\partial^2 W}{\partial \phi\partial p_2}+\frac{\partial^2 W}{\partial b_c\partial p_2}\right)+2p_2\frac{\partial^2 W}{\partial \phi\partial p_1}\right.$$

$$\left.+p_1p_2\frac{\partial^2 W}{\partial b_c\partial p_1}+p_{12}^2\frac{\partial^2 W}{\partial p_2^2}+2p_{12}p_2\frac{\partial^2 W}{\partial p_1\partial p_2}+p_2^2\frac{\partial^2 W}{\partial p_1^2}+p_2\frac{\partial W}{\partial b_c}\right]=0. \quad (3.22)$$

Equation (3.22) is required to be solved to obtain the characterstic funnction $W(r,\phi,b_c,p_1,p_2)$.

Since W is not a function of p_{22}, the coefficient of the term involving p_{22} should be zero. Hence, we have

$$\frac{\partial W}{\partial p_2}=o \quad , \quad (3.23)$$

from which we get

$$W = W(r,\phi,b_c,p_1)\,. \tag{3.24}$$

It is clear that W is not a function of p_{12}. Then the coefficients of the terms involving p_{12} and p_{12}^2 are equated to zero. By putting the coefficients of p_{12}^2 to zero, the characteristic function W is seen to be a linear function of p_1, i.e.

$$W = w_1(r,\phi,b_c)+p_1.w_2(r,\phi,b_c)\,. \tag{3.25}$$

Equating to zero the coefficient of p_{12} yields

$$\frac{\partial w_2}{\partial r}+p_2\frac{\partial w_2}{\partial b_c}=0\,. \tag{3.26}$$

Since w_2 is not a function of p_2, the cofficient of p_2 in (3.26) must vanish, which shows that w_2 is independent of b_c, i.e.

$$w_2 = w_2(r,\phi)\,. \tag{3.27}$$

The remaining term in equation (3.26) leads to the conclution that w_2 is independet of r and, as a result the charecterstic funtion now takes the form

$$W = w_1(r,\phi,b_c)+p_1.w_2(\phi)\,. \tag{3.28}$$

Putting this form of W into equation (3.22) and noting that both w_1 and w_2 are independent of p_1, equation (3.22) can be separated into three equations corresponding to the coefficients of p_1^0 , p_1^1 and p_1^2 . We then get:

Coefficient of p_1^2 :
$$\frac{\partial w_1}{\partial b_c}=0\,, \tag{3.29}$$

coefficient of p_1^1 :
$$r^2\frac{\partial w_1}{\partial b_c}+2\frac{\partial^2 w_1}{\partial b_c\partial\phi}+\frac{d^2 w_2}{d\phi^2}=0\,, \tag{3.30}$$

coefficient of p_1^0 : $r^2(\frac{\partial^2 w_1}{\partial r^2}+2p_2\frac{\partial^2 w_1}{\partial b_c \partial r}+p_2^2\frac{\partial^2 w_2}{\partial b_c^2})+r(\frac{\partial w_1}{\partial r}+p_2\frac{\partial w_1}{\partial b_c})$

$$+(\frac{\partial^2 w_1}{\partial \phi^2}+2p_2\frac{dw_2}{d\phi}+p_2\frac{\partial w_1}{\partial b_c})=0. \quad (3.31)$$

From (3.29), we have

$$w_1=w_{11}(r,\phi)+b_c.w_{12}(r,\phi) \quad (3.32)$$

Substituting by this form of w_1 into equation (3.30) yields

$$r^2 w_{12}+\frac{d^2 w_2}{d\phi^2}+2\frac{\partial w_{12}}{\partial \phi}=0. \quad (3.33)$$

It is clear that w_2 is independent of r. Hence, the coefficient of r^2 in (3.33) must vanish. This leads us to

$$w_{12}=0, \quad (3.34)$$

$$\frac{d^2 w_2}{d\phi^2}+2\frac{\partial w_{12}}{\partial \phi}=0. \quad (3.35)$$

Substituting by (3.34) into (3.35) and integrating, we get

$$w_2=C_0\phi+C_1, \quad (3.36)$$

where C_0 and C_1 are constants. Now, the function w_1 in (3.32) takes the form

$$w_1=w_{11}(r,\phi). \quad (3.37)$$

Since w_1 and w_2 are independent of p_2, then equation (3.31)can be separated into three equations corresponding to the coefficients of p_2^0, p_2^1 and p_2^2. We then get:

coefficient of p_2^2 : $\frac{\partial^2 w_1}{\partial b_c^2} = 0$, (3.38)

coefficient of p_2^1 : $2r^2 \frac{\partial^2 w_1}{\partial b_c \partial r} + \frac{dw_2}{d\phi} + (r+1)\frac{\partial w_1}{\partial b_c} = 0$, (3.39)

coefficient of p_2^0 : $r^2 \frac{\partial^2 w_1}{\partial r^2} + r\frac{\partial w_1}{\partial r} + \frac{\partial^2 w_1}{\partial \phi^2} = 0.$ (3.40)

Introduction of (3.36) and (3.37) into (3.38)-(3.40) yields

$$C_0 = 0, \quad (3.41)$$

$$r^2 \frac{\partial^2 w_{11}}{\partial r^2} + r\frac{\partial w_{11}}{\partial r} + \frac{\partial^2 w_{11}}{\partial \phi^2} = 0. \quad (3.42)$$

Equation (3.42)has the solution

$$w_{11} = (A_1 \cos k\phi + A_2 \sin k\phi)(a_1 r^k + a_2 r^{-k}), \quad (3.43)$$

where A_1, A_2, a_1, a_2 and k are constants.

Hence, the characteristic function W takes the form

$$W = (A_1 \cos k\phi + A_2 \sin k\phi)(a_1 r^k + a_2 r^{-k}) + C_1 p_1 . \quad (3.44)$$

To posses finite values at r =0 "since equation (2.19) is defined in $0 \le r \le 1$", we must take $a_2 = 0$. Hence, we have

$$W = C_1 p_1 - (C_2 \cos k\phi + C_3 \sin k\phi) r^k , \quad (3.45)$$

where C_1, C_2 and C_3 are constants.

The boundary condition at r=1, leads us to take k=1. Hence, the characteristic function W takes the form

$$W = C_1 p_1 - (C_2 \cos\phi + C_3 \sin\phi) r . \quad (3.46)$$

Applying equations (3.11)-(3.13), we get the following infinitesimals:

$$\zeta^{r}(r,\phi,b_c)=0\ , \tag{3.47}$$

$$\zeta^{\phi}(r,\phi,b_c)=C_1\ , \tag{3.48}$$

$$\eta^{b_c}(r,\phi,b_c)=r(C_2\cos\phi+C_3\sin\phi)\ . \tag{3.49}$$

3.3. Reduction of Independent Variables and Analytical Solution

In order to reduce (2.19) to an ordinary differential equation with only one independent variable, we construct similarity variable and similarity form of the field variable b_c. According to the well known method by Bluman and Cole [6], we have to solve the characteristic equations

$$\frac{dr}{\zeta^{r}}=\frac{d\phi}{\zeta^{\phi}}=\frac{db_c}{\eta^{b_c}}\ . \tag{3.50}$$

The general solution of (3.50) will involve two arbitrary constants, one of which takes the role of similarity variable $\eta=\eta(r,\phi)$ and the other, say F(η), plays the role of dependent variable, usually called the similarity function.

Substituting by (3.47)-(3.49) into (3.50) and solving, we have

$$\eta(r,\phi)=r\quad , \tag{3.51}$$

$$b_c(r,\phi)=F(\eta)+r[k_1\cos\phi+k_2\sin\phi]\quad , \tag{3.52}$$

where $k_1=C_3/C_1$ and $k_2=C_2/C_1$.

By substituting the similarity forms (3.51) and (3.52) in the partial differential equation (2.19), it will be reduced to the ordinary differential equation

$$\eta^2\frac{d^2F}{d\eta^2}+\eta\frac{dF}{d\eta}=0\quad , \tag{3.53}$$

which has the solution

$$F(\eta)=A+k_3\ln\eta\quad , \tag{3.54}$$

where A, k_3 are constants to be determined.

To posses finite value at r=0, the constant k_3 must vanish and hence, we have

$$b_c(r,\phi) = A + r[k_1 \cos\phi + k_2 \sin\phi] \quad . \tag{3.55}$$

Due to the symmetry of the problem with respect to the direction of the magnetic field; see[16],we have

$$v(x,y) = v(-x,y),\ b_f(x,y) = -b_f(-x,y),\ b_c(x,y) = -b_c(-x,y), \tag{3.56}$$

from which we have

$$b_c(r,\phi) = k_1 r.\cos\phi \quad . \tag{3.57}$$

4. Similarity Solutions for $B_F(R,\phi)$ and $V(R,\phi)$ when $M<<1$

In this section, we try to obtain the similarity solutions for the magnetic field perturbation $b_f(r,\phi)$ and the velocity $v(r,\phi)$.

Let

$$b_f(r,\phi) = \frac{1}{M}\psi_1(r,\phi) \quad , \tag{4.1}$$

$$v(r,\phi) = \psi_2(r,\phi) \quad .$$

Hence, equations (2.20) and (2.21) are

$$r^2\frac{\partial^2\psi_1}{\partial r^2} + r\frac{\partial\psi_1}{\partial r} + \frac{\partial^2\psi_1}{\partial r^2} + M^2\left(r^2\cos\phi\frac{\partial\psi_2}{\partial r} - r\sin\phi\frac{\partial\psi_2}{\partial\phi}\right) = 0 \quad , \tag{4.2}$$

$$r^2\frac{\partial^2\psi_2}{\partial r^2} + r\frac{\partial\psi_2}{\partial r} + \frac{\partial^2\psi_2}{\partial\phi^2} + r^2\cos\phi\frac{\partial\psi_1}{\partial r} - r\sin\phi\frac{\partial\psi_1}{\partial\phi} = 0 \quad , \tag{4.3}$$

where $1 < r < \infty$.

Using (3.57) and (4.1), the boundary conditions (2.22) and (2.23) are

$$\left.\begin{aligned}&\psi_2 = 0,\\ &\psi_1 = (k_1 M + 1)\cos\phi,\\ &\frac{\partial \psi_1}{\partial r} = \left(\frac{Mk_1}{\lambda} + 1\right)\cos\phi,\end{aligned}\right\} \text{ at } r=1, \tag{4.4}$$

$$\psi_2 \rightarrow 0 \ \psi_1 \rightarrow 0 \text{ as } r \rightarrow \infty. \tag{4.5}$$

4.1. Solution of Equation (4.2)

In case of weak magnetic field (M<<1), The terms multiplied by M^2 can be neglected . Hence, equation (4.2)reduces to the equation

$$r^2 \frac{\partial^2 \psi_1}{\partial r^2} + r\frac{\partial \psi_1}{\partial r} + \frac{\partial^2 \psi_1}{\partial r^2} = 0 \quad . \tag{4.6}$$

It is clear that equation (4.6) is identical to equation(2.19), but with ψ_1 instead of b_c. Applying the same procedure of section (3.2), we get the following characteristic function:

$$W_1 = (A_3 \cos k\phi + A_4 \sin k\phi)(a_3 r^k + a_4 r^{-k}) + C_4 q_1 \, . \tag{4.7}$$

where A_3, A_4, a_3, a_4, C_4 and k are constants.

To posses finite values at $r = \infty$ "since equation (4.2) is defined in $r > 1$", we must take $a_3 = 0$. Hence, we have

$$W_1 = C_4 q_1 - (C_5 \cos k\phi + C_6 \sin k\phi) r^{-k} \, , \tag{4.8}$$

where C_5 and C_6 are constants and

$$q_1 = \frac{\partial \psi_1}{\partial \phi}, \quad q_2 = \frac{\partial \psi_1}{\partial r}, \quad q_{11} = \frac{\partial^2 \psi_1}{\partial \phi^2}, \quad q_{22} = \frac{\partial^2 \psi_1}{\partial r^2} \text{ and } \quad q_{12} = \frac{\partial^2 \psi_1}{\partial r \partial \phi} \, . \tag{4.9}$$

The boundary condition at r =1 in (4.4) leads us to take k=1.Hence, characteristic function W_1 takes the form

$$W_1 = C_4 q_1 - \frac{1}{r}(C_5 \cos\phi + C_6 \sin\phi) \quad , \tag{4.10}$$

from which we obtain the following infinitesimals

$$\zeta_1^r(r, \phi, \psi_1) = \frac{\partial W_1}{\partial q_2} = 0 \ , \tag{4.11}$$

$$\zeta_1^\phi(r, \phi, \psi_1) = \frac{\partial W_1}{\partial q_1} = C_4 \quad , \tag{4.12}$$

$$\eta^{\psi_1}(r, \phi, \psi_1) = \frac{1}{r}(C_5 \cos\phi + C_6 \sin\phi) \ . \tag{4.13}$$

In the following, reduction of equation (4.2) is found with the aid of the infinitesimals ζ_1^r, ζ_1^ϕ and η^{ψ_1}.

Following the same procedure of section (3.3), similarity variable and similarity form of the field variable $\psi_1(r,\phi)$ are obtained by using the characteristic equations

$$\frac{dr}{\zeta_1^r} = \frac{d\phi}{\zeta_1^\phi} = \frac{d\psi_1}{\eta^{\psi_1}} \quad . \tag{4.14}$$

Introduction of (4.11)-(4.13) into (4.14) yields the similarity variable

$$\eta(r,\phi) = r \quad , \tag{4.15}$$

and the following similarity form of the field variable:

$$\psi_1(r,\phi) = g_1(\eta) + \frac{1}{r}(k_4 \cos\phi + k_5 \sin\phi) \quad , \tag{4.16}$$

where $k_4=-C_6/C_4$, $k_5=C_5/C_4$.

Substitution by (4.15) and (4.16) into equation (4.2) yields

$$\eta^2 \frac{d^2 g_1}{d\eta^2} + \eta \frac{dg_1}{d\eta} = 0 \ . \tag{4.17}$$

Solving (4.17) and using (4.16), we get

$$\psi_1(r,\phi) = k_6 + \frac{1}{r}(k_4 \cos\phi + k_5 \sin\phi) \ , \tag{4.18}$$

where k_6 is a constant to be determined. Applying the second part of the boundary condition (4.5), the second and the third parts of the boundary conditions (4.4), we get

$$k_5 = k_6 = 0, \tag{4.19}$$

$$k_1 = \frac{1}{M}\left(\frac{-2\lambda}{1+\lambda}\right) \ , \tag{4.20}$$

$$k_4 = \frac{1-\lambda}{1+\lambda} \ . \tag{4.21}$$

Hence, we have

$$\psi_1(r,\phi) = \left(\frac{1-\lambda}{1+\lambda}\right)\left(\frac{1}{r}\right)\cos\phi \ . \tag{4.22}$$

Using (4.1) and (3.57), we get

$$b_c(r,\phi) = \frac{1}{M}\left(\frac{-2\lambda}{1+\lambda}\right) r\cos\phi \ , \tag{4.23}$$

$$b_f(r,\phi) = \frac{1}{M}\left(\frac{1-\lambda}{1+\lambda}\right)\left(\frac{1}{r}\right)\cos\phi \ , \tag{4.24}$$

4.2. Solution of Equation (4.3)

Introduction of (4.22) into (4.3) yields

$$r^2\frac{\partial^2\psi_2}{\partial r^2}+r\frac{\partial\psi_2}{\partial r}+\frac{\partial^2\psi_2}{\partial\phi^2}-k_4\cos 2\phi=0 \ . \qquad (4.25)$$

Let us introduce the following notations

$$s_1=\frac{\partial\psi_2}{\partial\phi},\quad s_2=\frac{\partial\psi_2}{\partial r},\quad s_{11}=\frac{\partial^2\psi_2}{\partial\phi^2},\quad s_{22}=\frac{\partial^2\psi_2}{\partial r^2},$$

and $s_{12}=\dfrac{\partial^2\psi_2}{\partial r\partial\phi}$, (4.26)

where the characteristic function W_2 is defined as

$$W_2=\zeta_2^{\phi}s_1+\zeta_2^{r}s_2-\eta^{\psi_2}, \qquad (4.27)$$

where $\zeta_2^{\phi},\zeta_2^{r}$ and η^{ψ_2} are the infinitesimals of the group of transformations of equation(4.3)corresponding to ϕ, r and ψ_2; respectively, and are defined by

$$\zeta_2^{\phi}=\frac{\partial W_2}{\partial s_1},\ \zeta_2^{r}=\frac{\partial W_2}{\partial s_2} \text{ and } \eta^{\psi_2}=s_1\frac{\partial W_2}{\partial s_1}+s_2\frac{\partial W_2}{\partial s_2}-W_2. \qquad (4.28)$$

The invariance of equation (4.25) yields

$$2k_4\sin 2\phi\frac{\partial W_2}{\partial s_1}+(2rs_{22}+s_2)\frac{\partial W_2}{\partial s_2}-r(\frac{\partial W_2}{\partial r}+s_2\frac{\partial W_2}{\partial\psi_2})-$$

$$r^2\left[\frac{\partial^2 W_2}{\partial r^2}+2s_2\frac{\partial^2 W_2}{\partial r\partial\psi_2}+s_2^2\frac{\partial^2 W_2}{\partial\psi_2^2}+2s_{12}\left(\frac{\partial^2 W_2}{\partial r\partial s_1}+\frac{\partial^2 W_2}{\partial\psi_2\partial s_1}\right)+2s_1\frac{\partial^2 W_2}{\partial r\partial s_2}\right.$$

$$\left.+s_1s_2\frac{\partial^2 W_2}{\partial\psi_2\partial s_2}+s_{12}^2\frac{\partial^2 W_2}{\partial s_1^2}+2s_{12}s_1\frac{\partial^2 W_2}{\partial s_1\partial s_2}+s_1^2\frac{\partial^2 W_2}{\partial s_1^2}+s_1\frac{\partial W_2}{\partial\psi_2}\right]-$$

$$\left[\frac{\partial^2 W_2}{\partial\phi^2}+2s_1\frac{\partial^2 W_2}{\partial\phi\partial\psi_2}+s_1^2\frac{\partial^2 W_2}{\partial\psi_2^2}+2s_{12}\left(\frac{\partial^2 W_2}{\partial\phi\partial s_2}+\frac{\partial^2 W_2}{\partial\psi_2\partial s_2}\right)+2s_2\frac{\partial^2 W_2}{\partial\phi\partial s_1}\right.$$

$$+s_1 s_2 \frac{\partial^2 W_2}{\partial \psi_2 \partial s_1} + s_{12}^2 \frac{\partial^2 W_2}{\partial s_2^2} + 2 s_{12} s_2 \frac{\partial^2 W_2}{\partial s_1 \partial s_2} + s_2^2 \frac{\partial^2 W_2}{\partial s_1^2} + s_2 \frac{\partial W_2}{\partial \psi_2} \Bigg] = 0. \tag{4.29}$$

Following the same procedure of section (3.2), we get the characteristic function

$$W_2 = C_7 s_1 + \frac{1}{2} k_4 C_7 (\frac{1}{r^2} - 1) \sin 2\phi, \tag{4.30}$$

where C_7 is an arbitrary constant.

From (4.30) and using (4.28), we get the following infinitesimals:

$$\zeta_2^r(r, \phi, \psi_2) = 0 \ , \tag{4.31}$$

$$\zeta_2^\phi(r, \phi, \psi_2) = C_7 \ , \tag{4.32}$$

$$\eta^{\psi_2}(r, \phi, \psi_2) = \frac{1}{2} C_7 k_4 \left(1 - \frac{1}{r^2}\right)(\sin 2\phi) \ . \tag{4.33}$$

Following the same procedure of section (3.3), similarity variable and similarity form of the field variable $\psi_2(r, \phi)$ are obtained by using the characteristic equations

$$\frac{dr}{\zeta_2^r} = \frac{d\phi}{\zeta_2^\phi} = \frac{d\psi_2}{\eta^{\psi_2}} \ . \tag{4.34}$$

Introduction of (4.31)-(4.33) into (4.34) yields the similarity variable

$$\eta(r, \phi) = r \ , \tag{4.35}$$

and the following similarity form of the field variable:

$$\psi_2(r, \phi) = g_2(\eta) + \frac{1}{4} k_4 (\frac{1}{r^2} - 1) \cos 2\phi \ . \tag{4.36}$$

Under the similarity variable η in (4.35), the first part of the boundary conditions (4.4) is

$$g_2(1) = 0. \tag{4.37}$$

Substitution by (4.35) and (4.36) into equation (4.25) yields

$$\eta^2 \frac{d^2 g_2}{d\eta^2} + \eta \frac{dg_2}{d\eta} = 0 \quad , \tag{4.38}$$

which has the solution

$$g_2(\eta) = k_7 + k_8 \ln \eta \, , \tag{4.39}$$

where k_7 and k_8 are constants to be determined.

Applying the boundary condition (4.37) and using both (4.36)and(4.21),we get

$$v(r,\phi) = \left(\frac{1}{4}\right)\left(\frac{1-\lambda}{1+\lambda}\right)\left(\frac{1}{r^2} - 1\right)\cos 2\phi \quad . \tag{4.40}$$

Although the solution (4.44) for v gives a non-zero velocity at infinity, but it is coincident with the solution obtained by Gerbeth et al.[16]using a perturbation solution of v as a power series in M << 1 which ceases to be valid at large distances from the cylinder.

4.3. Case of an Insulating Cylinder

Within an insulating body, the current, and consequently b_c, is zero. The boundary conditions are simply, see [16]:

$$v = 0 \quad , \quad b_f = M^{-1} \cos\phi \quad , \quad \text{at} \quad r = 1 \, . \tag{4.41}$$

Applying the same procedure of the two previous sections, we reach to equations (3.57),(4.16) and (4.36). Using the boundary conditions (4.41) and (4.5), we obtain

$$\left.\begin{array}{l} k_1 = k_5 = 0 \quad , \\ k_1 M + 1 = k_4 \quad , \quad k_4 = 1 \quad , \\ g_1(\eta) = g_2(\eta) = 0 \quad . \end{array}\right\} \tag{4.42}$$

Hence, we have

$$b_c(r,\phi) = 0 \quad , \tag{4.43}$$

$$b_f(r,\phi) = \frac{1}{M}\frac{1}{r}\cos\phi \quad , \tag{4.44}$$

$$v(r,\phi)=\frac{1}{4}\left(\frac{1}{r^2}-1\right)\cos 2\phi \ . \tag{4.45}$$

It is important to notice that the solutions (4.43)-(4.45) can be obtained directly from (4.23),(4.24)and(4.40) by putting $\lambda=0$.

5. SIMILARITY SOLUTIONS FOR $B_F(R,\phi)$ AND $V(R,\phi)$ WHEN $M>>1$

In this section, we try to obtain the similarity solutions for the magnetic field perturbation $b_f(r,\phi)$ and the velocity $v(r,\phi)$.

Let

$$\psi_1(r,\phi)=e^{m(r-1)}\{b_f(r,\phi)+v(r,\phi)\} \ , \tag{5.1}$$

$$\psi_2(r,\phi)=e^{m(r-1)}\{b_f(r,\phi)-v(r,\phi)\} \ , \tag{5.2}$$

where m=M /N, N is a positive real number and its proper value will be determined later on . Hence, equations (2.20) and (2.21) are

$$r^2\frac{\partial^2\psi_1}{\partial r^2}+r(1-2\frac{M}{N}r)\frac{\partial\psi_1}{\partial r}+\frac{\partial^2\psi_1}{\partial\phi^2}+\frac{M}{N}r\psi_1(\frac{M}{N}r-1)$$

$$+M\left(r^2\cos\phi(\frac{\partial\psi_1}{\partial r}-\frac{M}{N}\psi_1)-r\sin\phi\frac{\partial\psi_1}{\partial\phi}\right)=0 \ , \tag{5.3}$$

$$r^2\frac{\partial^2\psi_2}{\partial r^2}+r(1-2\frac{M}{N}r)\frac{\partial\psi_2}{\partial r}+\frac{\partial^2\psi_2}{\partial\phi^2}+\frac{M}{N}r\psi_2(\frac{M}{N}r-1)$$

$$-M\left(r^2\cos\phi(\frac{\partial\psi_2}{\partial r}-\frac{M}{N}\psi_2)-r\sin\phi\frac{\partial\psi_2}{\partial\phi}\right)=0 \ , \tag{5.4}$$

where $1<r<\infty$.

Using (3.57), (5.1) and (5.2), the boundary conditions (2.22) and (2.23) are

$$\left.\begin{aligned}&\psi_1(r,\phi)=\psi_2(r,\phi),\\&\psi_1(r,\phi)=(k_1+M^{-1})\cos\phi,\\&\frac{\partial\psi_1}{\partial r}+\frac{\partial\psi_2}{\partial r}=2\left(k_1\left[\frac{1}{\lambda}+\frac{M}{N}\right]+\left[\frac{1}{N}+\frac{1}{M}\right]\right)\cos\phi,\end{aligned}\right\}\quad \text{at } r=1. \tag{5.5}$$

Following the same procedure in section (3.2), the invariance of equations (5.3) and (5.4) yields

$$r^2\left[\frac{\partial^2 W_1}{\partial r^2}+2q_2\frac{\partial^2 W_1}{\partial r\partial\psi_1}+q_2^2\frac{\partial^2 W_1}{\partial\psi_1^2}+2q_{12}\left(\frac{\partial^2 W_1}{\partial r\partial q_1}+\frac{\partial^2 W_1}{\partial\psi_1\partial q_1}\right)+2q_1\frac{\partial^2 W_1}{\partial r\partial q_2}\right.$$

$$\left.+q_1q_2\frac{\partial^2 W_1}{\partial\psi_1\partial q_2}+q_{12}^2\frac{\partial^2 W_1}{\partial q_1^2}+2q_{12}q_1\frac{\partial^2 W_1}{\partial q_1\partial q_2}+q_1^2\frac{\partial^2 W_1}{\partial q_1^2}+q_1\frac{\partial W_1}{\partial\psi_1}\right]+$$

$$\left[\frac{\partial^2 W_1}{\partial\phi^2}+2q_1\frac{\partial^2 W_1}{\partial\phi\partial\psi_1}+q_1^2\frac{\partial^2 W_1}{\partial\psi_1^2}+2q_{12}\left(\frac{\partial^2 W_1}{\partial\phi\partial q_2}+\frac{\partial^2 W_1}{\partial\psi_1\partial q_2}\right)+2q_2\frac{\partial^2 W_1}{\partial\phi\partial q_1}\right.$$

$$\left.+q_1q_2\frac{\partial^2 W_1}{\partial\psi_1\partial q_1}+q_{12}^2\frac{\partial^2 W_1}{\partial q_2^2}+2q_{12}q_2\frac{\partial^2 W_1}{\partial q_1\partial q_2}+q_2^2\frac{\partial^2 W_1}{\partial q_1^2}+q_2\frac{\partial W_1}{\partial\psi_1}\right]+$$

$$M\left[r^2\sin\phi\left(q_2-\frac{M}{N}\psi_1\right)+rq_1\cos\phi\right]\left(\frac{\partial\psi_1}{\partial q_1}\right)-Mr\sin\phi\left[\frac{\partial W_1}{\partial\phi}+q_1\frac{\partial W_1}{\partial\psi_1}\right]-$$

$$\left[2rq_{22}+\left(1-4r\frac{M}{N}\right)q_2+\psi_1(2r-1)\frac{M}{N}+M\left(2r\cos\phi(q_2-\frac{M}{N}\psi_1)-q_1\sin\phi\right)\right]\left(\frac{\partial W_1}{\partial q_2}\right)$$

$$+\left[r\left(1-2r\frac{M}{N}\right)+Mr^2\cos\phi\right]\left(\frac{\partial W_1}{\partial r}+q_2\frac{\partial W_1}{\partial\psi_1}\right)-$$

$$\left[r\frac{M}{N}\left(r\frac{M}{N}-1\right)-r^2\frac{M^2}{N}\cos\phi\right]\left(q_1\frac{\partial W_1}{\partial q_1}+q_2\frac{\partial W_1}{\partial q_2}-W_1\right)=0 \qquad ,(5.6)$$

and

$$r^2\left[\frac{\partial^2 W_2}{\partial r^2}+2s_2\frac{\partial^2 W_2}{\partial r\partial\psi_2}+s_2^2\frac{\partial^2 W_2}{\partial\psi_2^2}+2s_{12}\left(\frac{\partial^2 W_2}{\partial r\partial s_1}+\frac{\partial^2 W_2}{\partial\psi_2\partial s_1}\right)+2s_1\frac{\partial^2 W_2}{\partial r\partial s_2}\right.$$

$$\left.+s_1s_2\frac{\partial^2 W_2}{\partial\psi_2\partial s_2}+s_{12}^2\frac{\partial^2 W_2}{\partial s_1^2}+2s_{12}s_1\frac{\partial^2 W_2}{\partial s_1\partial s_2}+s_1^2\frac{\partial^2 W_2}{\partial s_1^2}+s_1\frac{\partial W_2}{\partial\psi_2}\right]+$$

$$\left[\frac{\partial^2 W_2}{\partial\phi^2}+2s_1\frac{\partial^2 W_2}{\partial\phi\partial\psi_2}+s_1^2\frac{\partial^2 W_2}{\partial\psi_2^2}+2s_{12}\left(\frac{\partial^2 W_2}{\partial\phi\partial s_2}+\frac{\partial^2 W_2}{\partial\psi_2\partial s_2}\right)+2s_2\frac{\partial^2 W_2}{\partial\phi\partial s_1}\right.$$

$$+s_1 s_2 \frac{\partial^2 W_2}{\partial \psi_2 \partial s_1} + s_{12}^2 \frac{\partial^2 W_2}{\partial s_2^2} + 2s_{12}s_2 \frac{\partial^2 W_2}{\partial s_1 \partial s_2} + s_2^2 \frac{\partial^2 W_2}{\partial s_1^2} + s_2 \frac{\partial W_2}{\partial \psi_2}\Bigg] -$$

$$M\left[r^2 \sin\phi\left(s_2 - \frac{M}{N}\psi_2\right) + rs_1 \cos\phi\right]\left(\frac{\partial \psi_2}{\partial s_1}\right) + Mr\sin\phi\left[\frac{\partial W_2}{\partial \phi} + s_1 \frac{\partial W_2}{\partial \psi_2}\right] -$$

$$\left[2rs_{22} + \left(1 - 4r\frac{M}{N}\right)s_2 + \psi_2(2r-1)\frac{M}{N} - M\left(2r\cos\phi(s_2 - \frac{M}{N}\psi_2) - s_1 \sin\phi\right)\right]\left(\frac{\partial W_2}{\partial s_2}\right) +$$

$$\left[r\left(1 - 2r\frac{M}{N}\right) - Mr^2 \cos\phi\right]\left(\frac{\partial W_2}{\partial r} + s_2 \frac{\partial W_2}{\partial \psi_2}\right) -$$

$$\left[r\frac{M}{N}\left(r\frac{M}{N} - 1\right) + r^2 \frac{M^2}{N}\cos\phi\right]\left(s_1 \frac{\partial W_2}{\partial s_1} + s_2 \frac{\partial W_2}{\partial s_2} - W_2\right) = 0 \qquad (5.7)$$

In case of strong magnetic field ($M >> 1$), the terms multiplied by "$1/M$" in the procedure can be neglected. Hence, following the same analysis as in section (3.2), we get the following characteristic functions:

$$W_1 = C_8(r\tan\phi)q_2 + C_9 q_1 + (3C_8 - C_9)\psi_1 \tan\phi, \qquad (5.8)$$

$$W_2 = C_{10}(r\tan\phi)s_2 + C_{11}s_1 + (3C_{10} - C_{11})\psi_2 \tan\phi, \qquad (5.9)$$

where C_8, C_9, C_{10} and C_{11} are arbitrary constants.

From (5.8) and (5.9), we get the following infinitesimals:

$$\zeta_1^r(r,\phi,\psi_1) = C_8 r\tan\phi \ , \qquad (5.10)$$

$$\zeta_1^\phi(r,\phi,\psi_1) = C_9 \ , \qquad (5.11)$$

$$\eta^{\psi_1}(r,\phi,\psi_1) = (C_9 - 3C_8)\psi_1 \tan\phi, \qquad (5.12)$$

and

$$\zeta_2^r(r,\phi,\psi_2) = C_{10} r\tan\phi \ , \qquad (5.13)$$

$$\zeta_2^\phi(r,\phi,\psi_2) = C_{11} \ , \qquad (5.14)$$

$$\eta^{\psi_2}(r,\phi,\psi_2) = (C_{11} - 3C_{10})\psi_2 \tan\phi. \tag{5.15}$$

Let

$$C_8 = C_{10} \text{ and } C_9 = C_{11}. \tag{5.16}$$

Hence, similarity variable and similarity forms of the field variables ψ_1 (r,ϕ) and ψ_2 (r, ϕ) are obtained by using the following equations:

$$\frac{dr}{C_8 r \tan\phi} = \frac{d\phi}{C_9} = \frac{d\psi_1}{(C_9 - 3C_8)\psi_1 \tan\phi} = \frac{d\psi_2}{(C_9 - 3C_8)\psi_2 \tan\phi}. \tag{5.17}$$

Solving (5.17) yields the similarity variable

$$\eta(r,\phi) = m r^k \cos\phi \ , \tag{5.18}$$

where $k=C_9/C_8$..

In general, the similarity forms of the dependent variables can be found by integrating one of the two equations

$$\frac{dr}{C_8 r \tan\phi} = \frac{d\psi_1}{(C_9 - 3C_8)\psi_1 \tan\phi} = \frac{d\psi_2}{(C_9 - 3C_8)\psi_2 \tan\phi} , \tag{5.19}$$

$$\frac{d\phi}{C_9} = \frac{d\psi_1}{(C_9 - 3C_8)\psi_1 \tan\phi} = \frac{d\psi_2}{(C_9 - 3C_8)\psi_2 \tan\phi}. \tag{5.20}$$

First approach: integrating (5.19) yields the following similarity forms of the field variables:

$$\psi_1(r,\phi) = \frac{1}{r^{3-k}} F_1(\eta) \ , \tag{5.21}$$

$$\psi_2(r,\phi) = \frac{1}{r^{3-k}} F_2(\eta) \ . \tag{5.22}$$

Substitution by (5.18), (5.21) and (5.22) into equations (5.3) and (5.4) yields, after neglecting the terms multiplied by "1/M", the following equations:

$$\frac{d^2F_1}{d\eta^2}+2(1-\cos\phi)\frac{dF_1}{d\eta}+(1-2\cos\phi)F_1=0 \quad , \tag{5.23}$$

$$\frac{d^2F_2}{d\eta^2}-2(1-\cos\phi)\frac{dF_2}{d\eta}+(1-2\cos\phi)F_2=0 \quad , \tag{5.24}$$

and

$$k=1 \ , N=2. \tag{5.25}$$

For equations (5.23) and (5.24) to be reduced to ordinary differential equations in a single independent invariant η, the coefficients should be constants or functions of η alone. Hence, we have

$$\frac{d^2F_1}{d\eta^2}+2\frac{dF_1}{d\eta}+F_1=0 \quad , \tag{5.26}$$

$$\frac{dF_1}{d\eta}+F_1=0 \quad , \tag{5.27}$$

$$\frac{d^2F_2}{d\eta^2}-2\frac{dF_2}{d\eta}+F_2=0 \quad , \tag{5.28}$$

$$\frac{dF_2}{d\eta}-F_2=0 \quad . \tag{5.29}$$

Solving equations (5.26)-(5.29) and using (5.21) and (5.22), we get

$$\psi_1(r,\phi)=\frac{k_9}{r^2}e^{-mr\cos\phi} \quad , \tag{5.30.1}$$

$$\psi_2(r,\phi)=\frac{k_{10}}{r^2}e^{mr\cos\phi} \quad , \tag{5.30.2}$$

where k^9 and k^{10} are constants to be determined with the aid of (5.5) .

Second approach: integrating (5.20) yields the following similarity forms of the field variables:

$$\psi_1(r,\phi) = (\cos\phi)^{\frac{3-k}{k}}\ \theta_1(\eta), \tag{5.31}$$

$$\psi_2(r,\phi) = (\cos\phi)^{\frac{3-k}{k}}\ \theta_2(\eta). \tag{5.32}$$

Substitution by (5.18), (5.31) and (5.32) into equations (5.3) and (5.4) and following the same procedure in first approach, we have

$$\psi_1(r,\phi) = k_{11}\,(\cos^2\phi)e^{-mr\cos\phi}\quad , \tag{5.33.1}$$

$$\psi_2(r,\phi) = k_{12}(\,\cos^2\phi)e^{mr\cos\phi}\quad , \tag{5.33.2}$$

where k^{11} and k^{12} are constants to be determined with the aid of (5.5).

The two solutions (5.30) and (5.33) are not identical. This may be due to neglecting the terms multiplied by "1/M". Also, they are not satisfying the boundary conditions (5.5).An approximate analytical solution can be obtained by adding or multiplying the solutions (5.30) and (5.33). In fact, adding them will result in a solution which does not satisfy the conditions (5.5). On the other hand, multiplying (5.30) and (5.33) yields, after replacing "r" by "r-1"in the exponential function (for compatibility with the boundary conditions (5.5)), the following:

$$\psi_1(r,\phi) = k_9\,\frac{\cos\phi}{r}\,e^{-m(r-1)\cos\phi}\quad , \tag{5.34}$$

$$\psi_2(r,\phi) = k_{10}\,\frac{\cos\phi}{r}\,e^{m(r-1)\cos\phi}\quad , \tag{5.35}$$

where $k^9 = k^{11}$ and $k^{10} = k^{12}$.

Hence, applying the boundary conditions (5.5) and using (3.57),(5.1) and (5.2), we have

$$b_c(r,\phi) = \frac{-\lambda(M+4)}{2M+\lambda M^2+2\lambda M}\,r\cos\phi\ , \tag{5.36}$$

$$b_f(r,\phi) = \frac{2(1-\lambda)}{2M+\lambda M^2+2\lambda M}\ e^{-m(r-1)}\ \frac{\cos\phi}{r}\ \cosh[m(r-1)\cos\phi]\ , \tag{5.37}$$

$$v(r,\phi) = \frac{2(\lambda-1)}{2M+\lambda M^2+2\lambda M}\ e^{-m(r-1)}\ \frac{\cos\phi}{r}\ \sinh[m(r-1)\cos\phi]\ . \tag{5.38}$$

Special case of an insulating cylinder

As mentioned before in section (4.3), the solutions corresponding to the case of an insulating cylinder can be obtained directly from (5.36)-(5.38) by putting $\lambda=0$. Hence, we have

$$b_c(r,\phi) = 0 \ , \tag{5.39}$$

$$b_f(r,\phi) = \frac{1}{M} e^{-m(r-1)} \frac{\cos\phi}{r} \cosh[m(r-1)\cos\phi] \ , \tag{5.40}$$

$$v(r,\phi) = -\frac{1}{M} e^{-m(r-1)} \frac{\cos\phi}{r} \sinh[m(r-1)\cos\phi] \ . \tag{5.41}$$

6. Discussion of the Physical Quantities and Concluding Remarks

6.1. Flow field

The dimensionless conductivity λ of the cylinder has a significant influence on the character of the flow. Observing (4.40) and (5.38), we find that in the case $\lambda<1$ (bad conductor) and $\lambda>1$ (good conducting cylinder) the velocities are of opposite sign. Also, for the trivial case ($\lambda=1$), where the conductivities σ_f and σ_c are equal, the solutions (4.40) and (5.38) refer to a zero velocity which meets the fact that the effective current passes the body without any perturbation.

It is important to notice that the characteristic functions W_1 and W_2 in section (5) is not defined at $\phi=\pm\pi/2$. Hence, the solutions presented in section (5) are valid in the vicinity of the cylinder, except for the tangential layers at $\phi=\pm\pi/2$ which can not be obtained by these formulas. Similar concluding remark for this point is found in [16].

6.2. Distribution of the Electric Current

The question arises, from the electro-dynamical point of view, as to how the electric current density is modified by the flow field. The main result is that the flow tends to homogenize the current. This is clear because the induced part of J is connected with v via

$$J_{ind} \sim V\times B, \tag{6.1}$$

and V is, in a crude approximation, proportional to $-(J_1)_y$, thus

$(J_{ind})_y \sim -(J_1)_y$, (6.2)

see Gerbth et al. [16].

The question now is: how does the value of the parameters effect on the relation between the homogeneous current within the cylinder and the incident current J? The solutions (4.23), (5.36) equation (2.4) and equation (2.7) give us the answer of this question as follows:

Introduction of (4.23) into (2.7) and using the reference values yields

$$\mathbf{B} = B_0\mathbf{e_x} - \frac{2\lambda}{1+\lambda} u_0 J_0 x \mathbf{e_z} \qquad (6.3)$$

Substitution by (6.3) into (2.4) yields the following relation between the homogeneous current within the cylinder and the incident current J:

$$J_c = \frac{2\lambda}{1+\lambda} J_0 \quad ; M<<1. \qquad (6.4.1)$$

In a similar manner, we have

$$J_c = \frac{\lambda(M+4)}{2\lambda+\lambda M+2} J_0 \approx \frac{\lambda M}{\lambda M+2} J_0 \quad ; M>>1. \qquad (6.4.2)$$

It is important to notice that the solutions (6.4) are coincident with the result obtained by [16] .Observing (6.4.2), we find that $J_c(\lambda, M) \to J_0$ as $M \to \infty$ except for $\lambda=0$ ($J_c=0$).It should be noted that for intermediate Hartmann numbers, J_c is not constant when $M>>1$.

6.3. Forces

6.3.1. Frictional Force

The dimensionless frictional force f_f can be obtained using the relation

$$f_f = \frac{1}{\pi}\int_0^{2\pi} \left.\frac{\partial v}{\partial r}\right|_{r=1} d\phi \quad , \qquad (6.5)$$

Inserting (4.40) and (5.38); respectively, into (6.5) and integrating, we get

$$f_f = 0 \; ; M<<1, \qquad (6.6.1)$$

$$f_f = \frac{\lambda-1}{2\lambda+\lambda M+2} \; ; M>>1. \qquad (6.6.2)$$

The solution (6.6.2) refers to that the dimensionless frictional force f_f (for M>>1) is mainly negative for $\lambda < 1$ and positive for $\lambda > 1$; respectively, see Figure 2.

6.3.2. Electromagnetic Force

The dimensionless electromagnetic force f_{em} is calculated using the relation

$$f_{em} = \frac{M}{\pi}\int_0^{2\pi}\left(b_c(r,\phi)\right)\Big|_{r=1} \cos\phi \, d\phi \quad (6.7)$$

Substitution by the obtained formularies for $b_c(r,\phi)$ into (6.7) and integrating, we have

$$f_{em} = -\frac{2\lambda}{1+\lambda}\ ; M \ll 1, \quad (6.8.1)$$

$$f_{em} = -\frac{\lambda(M+4)}{2\lambda+\lambda M+2}\ ; M \gg 1. \quad (6.8.2)$$

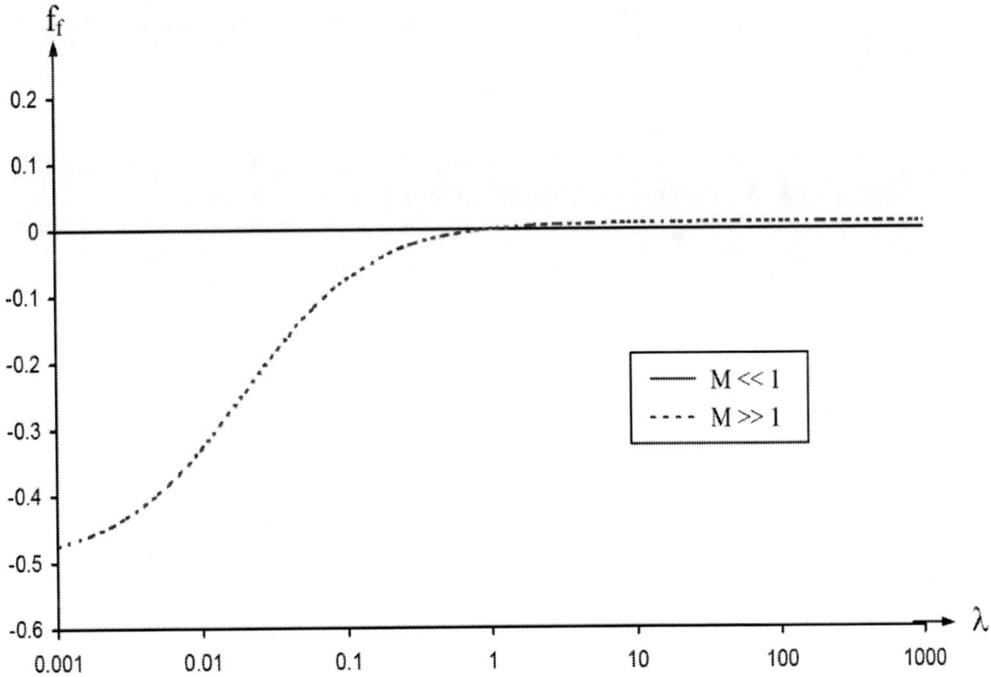

Figure 2. Frictional force f_f per unit length of a cylinder of infinite length.

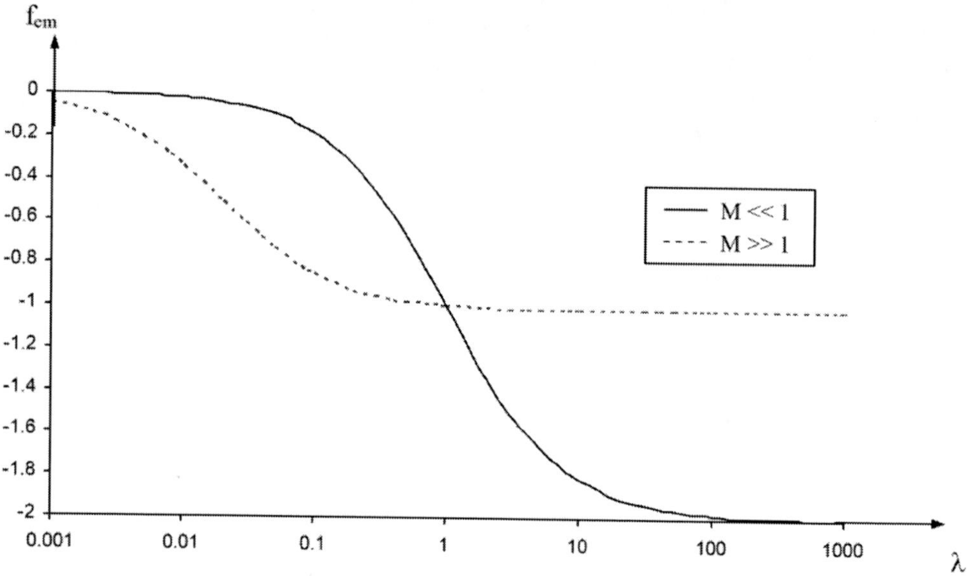

Figure 3. electromagnetic force f_{em} per unit length of a cylinder of infinite length.

The solutions (6.8) refer to that the dimensionless electromagnetic force f_{em} is always negative except for $\lambda=0$ ($f_{em}=0$). Figure 3 shows f_{em} calculated according to (6.8).From (6.8), we conclude that both the Hartmann number M and the dimensionless conductivity λ have an effect on the electromagnetic force f_{em} for $M>>1$ whereas the parameter λ only has an effect when $M<<1$.

6.3.3. Pressure and Total Force

Regarding the cylinder as an obstacle of finite length L, an additional force F_p due to the pressure difference between the two ends of the cylinder would be observed:

$$F_p = J_0 B_0 \pi a^2 L f_p \quad , \qquad f_p = +1. \tag{6.9}$$

Hence, we have the total force on the cylinder as

$$f(M,\lambda) = f_f + f_{em} + 1. \tag{6.10}$$

Substitution by (6.6) and (6.8) into (6.10) yields

$$f(M,\lambda) = \frac{1-\lambda}{1+\lambda} \quad ; M<<1, \tag{6.11.1}$$

$$f(M,\lambda) = \frac{1-\lambda}{2\lambda+\lambda M+2} \quad ; M>>1. \tag{6.11.2}$$

The solution (6.11.1) is coincident with the solutions obtained by [16] and [21] while (6.11.2) is believed to be new .The obtained results are plotted in Figure 4. From (6.11), the direction of the force depends on which of the following is realized:

a) $\lambda<1$, f is positive, reflecting that the main contribution comes from the pressure;
b) $\lambda=1$, the current is not perturbed, f is zero for all M;
c) $\lambda>1$, the influence of the electromagnetic force dominates, f is negative.

For the exceptional case of an insulating cylinder ($\lambda=0$), the solutions (6.11) predicts f=1 for M <<1 and f=1/2 for M>>1.

In a forthcoming paper, the direct method of Clarkson and Kruskal will be used in obtaining new similarity solutions for the problem considered here.

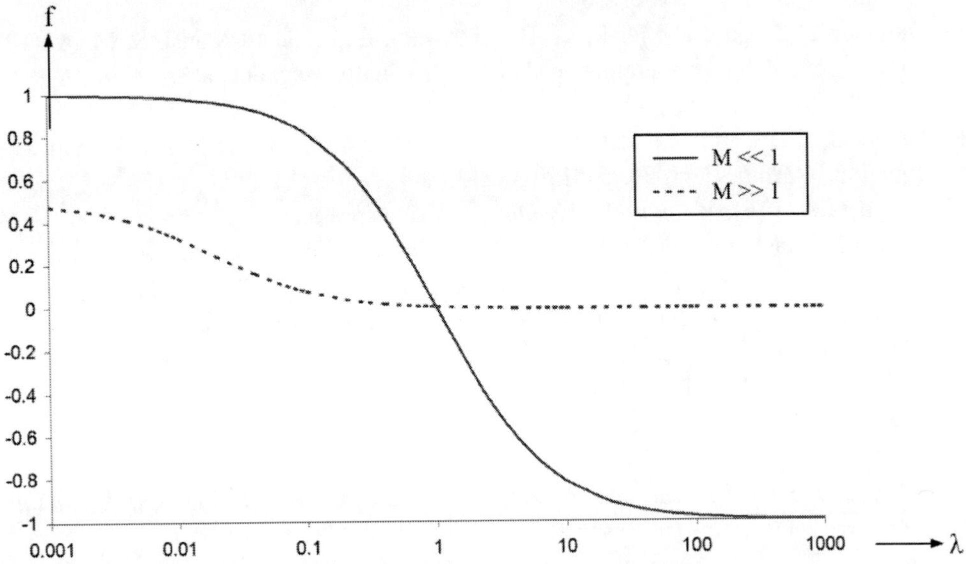

Figure 4. Total force f on a cylinder of finite length.

ACKNOWLEDGMENTS

The authors would like to express their gratitude to the American University in Cairo for offering a research grant to support this research.

REFERENCES

[1] Abd-el-Malek, M. B. and Badran, N. A. (1990), Group method analysis of unsteady free-convective laminar boundary-layer flow on a nonisothermal vertical circular cylinder, *Acta Mech.* 25(3-4), pp. 193-206.

[2] Abd-el-Malek, M. B. and Badran, N. A. (1991), Group method analysis of steady free-convective laminar boundary-layer flow on a nonisothermal vertical circular cylinder, *J. comput. Appl. Math.* 36(2), pp. 227-238.

[3] Abd-el-Malek, M. B. and El-Mansi, S. M. A. (2000), Group theoretic methods applied to Burgers' equation, *J. comput. Appl. Math.* 115(1-2), pp. 1-12.

[4] Abd-el-Malek, M. B. and El-Mansi, S. M. A. (2006), Lie group method for solving the MHD flow around a cylinder in crossed electric and weak magnetic fields, *Int. J. of Appl. Math and Mech* 2(2) pp.

[5] Birkhoff, G., *Hydrodynamics* (1950), Princeton Univ. Press, Princeton, NJ.

[6] Bluman, G. W. and Cole, J. D. (1974), *Similarity methods for differential equations*, Berlin.

[7] Bluman, G. W. and Cole, J. D. (1969), The general similarity solution of heat equation, *J. Math. Mech.* 11, pp. 1025-1042.

[8] Boutros, Y. Z., Abd-el-Malek, M. B., El-Awadi, I. A. and El-Mansi, S. M. A. (1991), Group method for temperature analysis of thermally stagnant lakes, *Acta Mech.* 133, pp. 131-144.

[9] Clarkson, P. A. and Kruskal, M. D. (1989), New similarity reduction of Boussinesq equation, *J. Math. Phys.* 30, pp. 2201-2213.

[10] Dragos, L. (1975), *Magnetofluiddynamics*, Abacus press.

[11] Fuchs, J. C. and Richter, E. W. (1987), Similarity solutions for the two - dimensional non- stationary ideal MHD equations, *J. Phys. A, Math. Gen.* 20, pp. 3135-3157.

[12] Fuchs, J. C. (1991), Symmetry groups and similarity solutions of MHD equations, *J. Math. Phys.* 32, pp. 1703 - 1708.

[13] Gailities, A. and Gerbeth, G. (1987), On stokes flow around a circular cylinder in a strong transverse magnetic fields (in Russion), *Magnitana Gidrodinamika,* pp. 137 – 139.

[14] Gelfgat, J. M., Lilausis, D. A. and Sherbinin, E. V. (1976), *Liquid metals in electromagnetic fields* (in Russian), Riga.

[15] Gerbeth, G. (1987), *Contributions to the theory of MHD two - phase flow*, Ph. D. thesis, Acad. Sci., G. D. R.

[16] Gerbeth, G., Thess, A. and Marty, P. (1990), Theoretical study of the MHD flow around a cylinder in crossed electric and magnetic fields, *Eur. J. Mech. B / Fluids.* 9, 239-257.

[17] Hunt, J. C. R. and Ludford, G. S. S. (1968), Three- dimensional MHD duct flows with strong transverse magnetic fields, *J. fluid Mech.* 33, pp. 693-714.

[18] Hunt, J. C. R. and Williams, W. E. (1968), Some electrically driven flows, J. fluid Mech. 31,705 - 722; *J. fluid Mech.* 33, 775-801.

[19] Leenov, D. and Kolin, A. (1954), Theory of electromagnetophoresis, *J. Chem. Phys.* 22, pp. 683-688.

[20] Marty, P. and Alemany (1982), A., *Theoretical and experimental aspects of electromagnetic separation*, IUTAM Symposium on Metallurgical Applications of MHD, Cambridge. Pp. 245- 259.

[21] Marty, P. and Alemany, A. (1983), Êcoulement dû a des champs magnêtique et êlectrique croisês autour d' un cylinder de conductivitê quelconque, *J. Mêc. Thêor. Appl.* 2, pp. 227-243.

[22] Na, T. Y. and Hansen, A. G. (1971), *Similarity analysis of differential equations by Lie group*, Journal of Franklin Institute. 6 .

[23] Noel. C. L. (1967), *Magnetohydrodynamics*, D. Van Nostrand Company, Inc.

[24] Olver, P. J. (1986), *Application of Lie groups to differential equations*, Berlin.

[25] Ovsiannikov, L. V. (1982), *Group analysis of differential equations,* New York.

[26] Ricou, R. (1975), Perturbations des paramet êlectriques et mêcaniques dues a l'introduction d'un porte - sonde au sein d'un fluide êlectroconducteur, en presence d'un champ de force êlectromagnêtique, Thêse de 3^{e} cycle, Universitê d'Aix – Marseille - III .

[27] Rogers, C. and Ames, W. F. (1989), *Nonlinear boundary value problems in science and engineering,* Academic Press, Inc., London.

[28] Rogers, C., Stallybrass, M. P. and Clement. D. L. (1983), On two phase filtration under gravity and with boundary infiltration: Application of Bäcklund transformation, Nonlinear Anal. *Theory, Methods Appl.* 7, pp. 785-799.

[29] Schwarz, F. (1988), Symmetries of differential equations; from Sophus Lie to computer algebra, *SIAM Review* 30, pp. 450-481.

[30] Seshadri, R. and Na, T. Y. (1985), *Group invariance in engineering boundary value problems,* Springer-Verlag, New York.

[31] Waechter, R. T. (1968), Steady longitudinal motion of an insulating cylinder in a conducting fluid, *Proc. Camb. Phil. Soc.* 64, pp. 1165-1201.

In: New Developments in Magnetohydrodynamics … ISBN: 978-1-61942-078-6
Editors: Stephen Talton and Patrick F. Theba 2012 Nova Science Publishers, Inc.

Chapter 6

A UNIFIED APPROACH TO ANALYTICAL SOLUTION OF AN UNSTEADY MHD GENERALIZED COUETTE FLOW WITH TRANSPIRATION

Basant K. Jha *** and Haruna M. Jibril***
Department of Mathematics, Faculty of Science,
Ahmadu Bello University, Zaria,
Federal Republic of Nigeria

ABSTRACT

The mathematical model describing the unsteady generalized Couette flow with porous boundaries of viscous incompressible, electrically conducting fluid is solved analytically for general nature of time-dependent pressure gradient and initial condition. For the general nature of motion of upper boundary, unified expression has been derived for the fluid velocity corresponding to the cases of magnetic field fixed relative to the fluid or to the upper boundary.

1. INTRODUCTION

Couette flow or pressure driven flow (Poiseuille flow) are well known problem in the history of fluid mechanics [1-6].This problem is of fundamental importance as it provides the exact solution of the Navier-Stokes equation and reveals how the velocity profile varies with time, approaching a linear velocity distribution asymptotically, and how the boundary layer spreads through the flow field. The steady and unsteady Couette-type flow in hydrodynamics and magneto-hydrodynamics, subjected to wall suction have been discussed by several authors [7-13], due to their applications in transpiration cooling and gaseous diffusion processes. Unsteady flow in porous channel, in particular is of utmost importance in the dialysis of blood in artificial kidneys. Wadhwa [14] studied the magneto-hydrodynamic

* e-mail: basant777@yahoo.co.uk; e-mail alharun2004@yahoo.com

generalized Couette flow by considering suction/injection at the plates, time-dependent pressure gradient, and arbitrary initial velocity. Wadhwa[14] presented their analysis by taking the magnetic lines of force fixed relative to the fluid. In this work the same problem is considered when the magnetic lines of force are fixed relative to the moving plate or fluid. \

2. Formulation of the Problem and Solution

The physical situation considered is that of the time dependent hydro-magnetic flow of a viscous, incompressible and electrically conducting fluid bounded by two infinite parallel porous plates, distant h apart, in which the bottom plate is fixed and subject to a mass injection constant velocity V_0 and there is a mass suction velocity V_0 at the top plate. The $x-$ axis is chosen along the lower porous plate in the horizontal direction and the $y-$axis normal to it. Initially, $(t \leq 0)$,both the plates are assumed to be at rest and the fluid velocity is $U_0(y)$.At t >0, the upper plate start moving with a velocity $U(t)$and a time dependent pressure gradient is applied along the plate direction. It is assumed that the plates are electrically non-conducting and applied uniform magnetic field B_0 is acting perpendicular to the flow direction. Since the porous plates are infinite in extent all the variable quantities are taken to be function of y and t only. By assuming a very small magnetic Reynolds number the induced magnetic field is neglected[15] in comparison to the applied magnetic field $(0, B_0, 0)$.Since no applied or polarization voltage is imposed on the flow field, the electric field vector E=0.This then corresponds to the case when no energy is added or extracted from the fluid by the electric field.

Within the framework of such assumptions the governing equation for the present problem can be obtained as

$$\rho\left[\frac{\partial u}{\partial t}+V_0\frac{\partial u}{\partial y}\right]=-\frac{\partial p}{\partial x}+\mu\frac{\partial^2 u}{\partial y^2}-\sigma B_0^2\left[u-KU(t)\right] \tag{1}$$

where

$$K=\begin{cases}0 \text{ if magnetic field is fixed relative to the fluid}\\ 1 \text{ if magnetic field is fixed relative to the plate}\end{cases}$$

$$u(0,t)=0;\ u(h,t)=U(t) \tag{2}$$

$$u(y,0)=U_0(y) \tag{3}$$

Let the solution be assumed in the form [4]

$$u(y,t)=\sum_{j=1}^{\infty} F_j(y)T_j(t)+\alpha(y)P(t)+\beta(y)U(t) \tag{4}$$

where

$$P(t)=-\frac{\partial p}{\partial x};\quad \alpha(y)=\sum_{j=1}^{\infty} a_j F_j(y);$$

$$\beta(y)=\sum_{j=1}^{\infty} b_j F_j(y);\quad U_0(y)=\sum_{j=1}^{\infty} c_j F_j(y).$$

Substituting (4) into equations (1)-(3) the original problem can be split in to the four sub-problems

$$\mu\frac{d^2F_j}{dy^2}-\rho V_0\frac{dF_j}{dy}-\sigma B_0^2 F_j=\lambda_j F_j \tag{5}$$

$$F_j(0)=F_j(h)=0,\quad j=1,2,3,\ldots\ldots\ldots\ldots$$

$$\mu\frac{d^2\alpha_j}{dy^2}-\rho V_0\frac{d\alpha_j}{dy}-\sigma B_0^2\alpha_j=-1, \tag{6}$$

$$\alpha_j(0)=\alpha_j(h)=0.$$

$$\mu\frac{d^2\beta_j}{dy^2}-\rho V_0\frac{d\beta_j}{dy}-\sigma B_0^2\beta_j=-K\sigma B_0^2,$$

$$\beta_j(0)=\beta_j(h)=1. \tag{7}$$

$$\rho\left[\frac{d\ T_j}{dt}+a_j\frac{dP}{dt}+b_j\frac{dU}{dt}\right]-\lambda_j T_j=0, \tag{8}$$

$$T_j(0)=c_j-a_jP(0)-b_jU(0).$$

Equation (5) is an eigen-value problem with eigen-values λ_j ; and equation (6) and (7) are non-homogeneous boundary-value problems. Solving (5) one gets

$$\lambda_j = -\frac{\mu}{h^2}\left[S^2 + M^2 + (j\pi)^2\right], \tag{9}$$

$$F_j(y) = Exp\left(\frac{Sy}{h}\right)Sin\left(\frac{j\pi y}{h}\right), \tag{10}$$

where

$S = \frac{\rho V_0 h}{2\mu}$ =suction Reynolds number

$M^2 = \frac{\sigma B_0^2 h^2}{\mu}$ =Hartmann number

It is clear from equation (10) that the eigen-functions do not depend on the Hartmann number.

The differential equation (5) can be written in the self-adjoint form

$$\frac{d}{dy}\left[Exp\left(-\frac{2Sy}{h}\right)\frac{dF_j}{dy}\right] + \left[\frac{\left(S^2 + (j\pi)^2\right)}{h^2}Exp\left(-\frac{2Sy}{h}\right)\right]F_j = 0.$$

Thus the eigen-functions $F_j(y)$ are orthogonal with respect to the weight-function $Exp\left(-\frac{2Sy}{h}\right)$, i.e.,

$$\int_0^h F_j(y)F_k(y)Exp\left(-\frac{2Sy}{h}\right)dy = 0, \quad \text{if } j \neq k$$

$$\int_0^h F_j^2(y)Exp\left(-\frac{2Sy}{h}\right)dy = \frac{h}{2}, \quad \text{if } j = k. \tag{11}$$

Therefore

$$a_j = \frac{2}{h}\int_0^h Exp\left(-\frac{2Sy}{h}\right)Sin\left(\frac{j\pi y}{h}\right)\alpha(y)dy = 0, \tag{12}$$

$$b_j = \frac{2}{h}\int_0^h Exp\left(-\frac{2Sy}{h}\right)Sin\left(\frac{j\pi y}{h}\right)\beta(y)dy = 0, \tag{13}$$

$$c_j = \frac{2}{h}\int_0^h Exp\left(-\frac{2Sy}{h}\right)Sin\left(\frac{j\pi y}{h}\right)U_0(y)dy = 0. \tag{14}$$

The solution of equation (6) and (7) are

$$\alpha(y) = \frac{h^2}{\mu M^2}\left[\left(1 - Cosh\left(\frac{Ny}{h}\right)Exp\left(S\frac{y}{h}\right)\right) + \left[(Cosh(S) - Exp(-S))\frac{Sinh\left(\frac{Ny}{h}\right)}{Sinh(N)}\right]Exp\left(S\frac{y}{h}\right)\right] \tag{15}$$

$$\beta(y) = \lfloor Exp(-S) + K(Cosh(N) - Exp(-S))\rfloor\frac{Sinh\left(\frac{Ny}{h}\right)}{Sinh(N)}Exp\left(S\frac{y}{h}\right)$$

$$+K\left[1. - Exp\left(S\frac{y}{h}\right)Cosh\left(\frac{Ny}{h}\right)\right] \tag{16}$$

where

$$N^2 = S^2 + M^2.$$

Using the expressions for $\alpha(y)$ and $\beta(y)$ from equation (15) and (16) in to equation (12) and (13), one gets

$$a_j = \frac{2j\pi h^2[1. - Exp(-S)Cos(j\pi)]}{\mu[S^2 + (j\pi)^2]n}, \tag{17}$$

$$b_j = -\frac{2j\pi Exp(-S)Cos(j\pi)}{n}. \tag{18}$$

Solution of equation (8) now becomes

$$T_j(t) = Exp\left(-\frac{\mu tn}{\rho h^2}\right)\left[T_j(0) - \int_0^t Exp\left(\frac{\mu tn}{\rho h^2}\right)\left\{a_j\frac{dP(t)}{dt} + b_j\frac{dU(t)}{dt}\right\}dt\right], \tag{19}$$

where

$$n=\left[S^2+M^2+(j\pi)^2\right].$$

Particular Cases

Several particular cases of physical interest can be derived from Equations (9),(10),and (15)-(19).

1. Generalized Couette Flow of Conduction Fluid in the Absence of Suction/Injection

In the absence of suction/injection $V_0=0$ so that $S=0$ .By doing the same, in equations (9),(10) and (15)-(19) the velocity distribution of an incompressible viscous electrically-conducting fluid due to time dependent pressure gradient, arbitrary initial velocity, arbitrary time-dependent motion of the upper plate, in the presence of transverse magnetic field fixed relative to the fluid or moving with upper plate, is

$$F_j(y)=Sin\left(\frac{j\pi y}{h}\right),$$

$$\alpha(y)=\frac{h^2}{\mu M^2}\left[1.-\frac{Sinh\left(\frac{My}{h}\right)+Sinh\left(M\left(1.-\left(\frac{y}{h}\right)\right)\right)}{Sinh(M)}\right],$$

$$\beta(y)=K\left(1-Cosh\left(\frac{My}{h}\right)\right)+\frac{Sinh\left(M\frac{y}{h}\right)}{Sinh(M)}\left[1.+K(Cosh(M)-1.)\right], \qquad (20)$$

$$a_j=\frac{2h^2\left[1.-Cos(j\pi)\right]}{\mu j\pi n_1}, b_j=-\frac{2j\pi Cos(j\pi)}{n_1},$$

$$T_j(t)=Exp\left(-\frac{\mu t n_1}{\rho h^2}\right)\left[T_j(0)-\int_0^t Exp\left(\frac{\mu t n_1}{\rho h^2}\right)\left\{a_j\frac{dP(t)}{dt}+b_j\frac{dU(t)}{dt}\right\}dt\right],$$

where

$$n_1=\left[M^2+(j\pi)^2\right].$$

Using the expression (20) in to equation (4), gives the required velocity distribution.

2. Generalized Couette Flow of Fluid in the Absence of Externally Applied Magnetic Field With Suction/Injection

In the absence of externally applied magnetic field with suction/injection $B_0 = 0$ so that $M = 0$ .By doing the same, in equations (9),(10) and (15)-(19) the velocity distribution of an incompressible viscous fluid due to time dependent pressure gradient, arbitrary initial velocity, arbitrary time-dependent motion of the upper plate, is

$$F_j(y) = Exp\left(\frac{Sy}{h}\right) Sin\left(\frac{j\pi y}{h}\right),$$

$$\alpha(y) = \frac{h^2}{\mu M^2}\left[\left(1 - Cosh\left(\frac{Sy}{h}\right) Exp\left(S\frac{y}{h}\right)\right) + \left[\left(Cosh(S) - Exp(-S)\right)\frac{Sinh\left(\frac{Sy}{h}\right)}{Sinh(S)}\right] Exp\left(S\frac{y}{h}\right)\right],$$

$$\beta(y) = \left[Exp(-S) + K\left(Cosh(S) - Exp(-S)\right)\right]\frac{Sinh\left(\frac{Sy}{h}\right)}{Sinh(S)} Exp\left(S\frac{y}{h}\right) \qquad (21)$$

$$+ K\left[1. - Exp\left(S\frac{y}{h}\right) Cosh\left(\frac{Sy}{h}\right)\right],$$

$$a_j = \frac{2j\pi h^2\left[1. - Exp(-S)Cos(j\pi)\right]}{\mu n_2^2}, b_j = -\frac{2j\pi Exp(-S)Cos(j\pi)}{n_2},$$

$$T_j(t) = Exp\left(-\frac{\mu t n_2}{\rho h^2}\right)\left[T_j(0) - \int_0^t Exp\left(\frac{\mu t n}{\rho h^2}\right)\left\{a_j\frac{dP(t)}{dt} + b_j\frac{dU(t)}{dt}\right\}dt\right],$$

where

$$n_2 = \left[S^2 + (j\pi)^2\right].$$

Substitution of expression (21) in equation (4) gives the desired velocity distribution.

3. Generalized Couette Flow of Fluid in the Absence of Externally Applied Magnetic Field and Suction/Injection

In the absence of externally applied magnetic field and suction/injection $B_0 = 0$ and $V_0 = 0$ so that $M = 0$ and $S = 0$..By doing the same, in equations (9),(10) and (15)-(19) the velocity distribution of an incompressible viscous fluid due to time dependent pressure gradient, arbitrary initial velocity, arbitrary time-dependent motion of the upper plate, is

$$F_j(y)=Sin\left(\frac{j\pi y}{h}\right),$$

$$\alpha(y)=\frac{h^2}{2\mu}\left[\frac{y}{h}-\left(\frac{y}{h}\right)^2\right],$$

$$\beta(y)=\frac{y}{h},$$

$$a_j=\frac{2j\pi h^2[1.-Cos(j\pi)]}{\mu n_3^2}, \tag{22}$$

$$b_j=-\frac{2j\pi Cos(j\pi)}{n_3},$$

$$T_j(t)=Exp\left(-\frac{\mu t n_3}{\rho h^2}\right)\left[T_j(0)-\int_0^t Exp\left(\frac{\mu t n_3}{\rho h^2}\right)\left\{a_j\frac{dP(t)}{dt}+b_j\frac{dU(t)}{dt}\right\}dt\right],$$

where

$$n_3=(j\pi)^2.$$

Using expression (22) in equation (4) gives the required velocity distribution.

Each solutions will be completed when the forms of $U(t), P(t)$ and $U_0(t)$ are properly defined.

REFERENCES

[1] H.Schlichting and K. Gersten, Boundary Layer Theory, 8th revised and enlarged edition (English), Springer- verlag, New ork (2000).

[2] F.M.White, Viscous Fluid Flow, second edition, p144, McGraw-Hill Company, New York (1991).

[3] R.L. Panton, Incompressible Flow, second edition, p.310, J. Wiley, New York (1996).

[4] P.A. Longwell, Mechanics of Fluid Flow, p.158, McGraw-Hill Company, New York (1966).

[5] D.F. Rogers, Laminar Flow Analysis, Cambridge University Press, New York (1992).

[6] V.N. Constantinescu, Laminar Viscous Flow, Springer-verlag, New York (1995).

[7] Tiegang Fang, A note on the incompressible Couette flow with porous walls, Int.

[8] Comm. Heat Mass Transfer, vol.31, p31 (2004).

[9] Tiegang Fang, Further discussion on the incompressible pressure-driven flow in a
[10] channel with porous walls, Int. Comm. Heat Mass Transfer, vol.31,p487(2004).
[11] K.R.Cramer, A Generalized porous-wall Couette-type flow, Jr. Aerospace
[12] Sci., vol.26,p121(1959).
[13] P.D.Verma., and J.N.Bansal, Flow of viscous incompressible fluid between two
[14] parallel plates, one in uniform motion andother at rest with uniform suction at the
[15] stationary plate, Proc. Indian Acad. Sci.,vol.64,p285(1966).
[16] P.Ramamoorty,A generalized porous wall Couette type flow with uniform suction
[17] or blowing and uniform transverse magnetic field, Jr. Aerospace Sci., vol.29, p1111962).
[18] R.K.Rathy, Hydromagnetic Couette's flow with suction and injection, ZAMP,
[19] vol.17,p370 (1963).
[20] P.K.Muhuri, Flow formation in Couette motion in magneto-hydro-dynamics with
[21] suction, Jr. Phy. Soc.Japan,vol. 18,p1671(1963).
[22] Y.D.Wadhwa, Magneto-hydro-dynamics Generalized Couette flow (With
[23] Uniform Suction, Time-Dependent Pressure Gradient: and Arbitrary Initial
[24] Velocity), Jr. Appl. Mechanics,p 1189(1970).
[25] G.W.Sutton., and A. Sherman, Engineering Magneto-hydro-dynamics, McGraw-
[26] Hill, New York,(1965).

In: New Developments in Magnetohydrodynamics ... ISBN: 978-1-61942-078-6
Editors: Stephen Talton and Patrick F. Theba 2012 Nova Science Publishers, Inc.

Chapter 7

UNSTEADY MHD TWO PHASE FLOW: A GREENS FUNCTION APPROACH

Basant K. Jha* and Haruna M. Jibril
Department of Mathematics, Faculty of Science, Ahmadu Bello University, Zaria, Federal Republic of Nigeria

ABSTRACT

A study is made to analyze the unsteady MHD (magneto-hydrodynamic) two phase flow in a horizontal channel filled with conducting fluids of different physical properties in the presence of transverse magnetic field for short circuit case. It is assumed that the conducting fluids are having constant transport properties. The two regions are coupled by equating the velocity and shear-stress at the interface.

Using the Green's function method, expressions for velocities for both phases are obtained for general class of time dependent movement of boundaries or sudden change in pressure gradient or both. As a special case, expressions for velocities in both phases are obtained due to sudden change in pressure gradient.

1. INTRODUCTION

The study of the dynamics of time dependent MHD (magneto hydrodynamic) two phase flows find applications to variety of problems in astrophysical and geophysical phenomena. When the combined effects of magnetic and fluid forces are incorporated into the governing equations of electrically conducting fluid flows, these equations become intractable for an analytical treatment. Accordingly, simplifications have been introduced in to the fluid models in literature with view to analyzing specific flow features. For instance, Shail [1] considered the two-phase flow of conducting and non-conducting incompressible fluids in a horizontal channel in the presence of transverse magnetic field. It was shown theoretically that significant increase in the flow rate of conducting fluid can be obtained for suitable choice of

* e-mail: basant777@yahoo.co.uk

ratio of viscosities and the depth of the two phases. Lohrasbi and Sahai[2] studied the thermal characteristics of [1],by accommodating the viscous and Ohmic-dissipation in the energy equation. The two- phase flows was further examined by Malashetty and Leela [3,4] by considering both phases as an electrically conducting and solved the momentum and energy equation equations analytically for short circuit and open circuit cases respectively.

In all these above discussed works momentum and energy equations are assumed to be independent of time. This work is devoted to study the time dependent MHD two phase flows of viscous, incompressible and electrically conducting fluids in a horizontal parallel plate channel having both phases occupied by electrically conducting fluids of different nature. An exact solution is obtained using Green's function approach.

2. Governing Equations and Solutions

The physical situation corresponds to that of the time dependent MHD two phase flows in a horizontal parallel plate channel filled with two different conducting fluids. The physical model is shown in fig.1. The direction of the flow is taken along the x'-axis coinciding with the lower horizontal plate while y'-axis is taken normal to the flow direction. The regions $0 \le y \le d$ and $d \le y \le h$ are occupied by two different conducting fluids having different densities, viscosities and electrical conductivities. An external magnetic field of uniform strength B_0, is applied parallel to y'-axis. The magnetic field is assumed to be small so that the induced magnetic field can be neglected compared to the applied magnetic field. The unsteadiness in the fluid motion is caused either by a sudden change in the imposed pressure gradient along x'-axis or (and) by a sudden change in the velocity of the channel boundaries.

Under the assumptions discussed above the one dimensional time dependent flow for both phases in short-circuit case are governed by the following non-dimensional equations:

$$\frac{\partial U_1}{\partial t} = -P + \frac{\partial^2 U_1}{\partial Y^2} - M^2 U_1 \tag{1}$$

$$\frac{\partial U_2}{\partial t} = -\alpha P + \gamma\alpha \frac{\partial^2 U_2}{\partial Y^2} - \alpha\delta M^2 U_2 \tag{2}$$

The Initial, boundary and interface conditions are:

$$\begin{aligned} &U_1(0,Y) = F_1(Y), U_2(0,Y) = F_2(Y), \\ &U_1(t,0) = f_1(t), \\ &U_2(t,H) = f_2(t), \end{aligned} \tag{3}$$

$$U_1(t,1) = U_2(t,1), \frac{\partial U_1(t,1)}{\partial Y} = \gamma \frac{\partial U_2(t,1)}{\partial Y}, \tag{4}$$

The dimensionless quantities introduced in equations (1-4) are given in list of symbols.

Equation (3) represents the general nature of initial and boundary conditions. A specific situation can be handled by proper selection of initial and boundary conditions.

It is more convenient to solve the problem with homogeneous boundary conditions. To attain this let

$$U_i(Y,t)=V_i(Y,t)+\phi_i(Y)f_1(t)+\varepsilon_i(Y)f_2(t), i=1,2 \quad (5)$$

This relation transforms the governing equations (1-4) in to the following set of equations

$$\frac{d^2\phi_1}{dY^2}-M^2\phi_1=0 \quad (6)$$

$$\gamma\frac{d^2\phi_1}{dY^2}-\delta M^2\phi_1=0 \quad (7)$$

$$\left.\begin{array}{l}\phi_1(0)=1;\phi_2(H)=0,\\ \phi_1(1)=\phi_2(1),\dfrac{d\phi_1(1)}{dY}=\gamma\dfrac{d\phi_2(1)}{dY}\end{array}\right\}. \quad (8)$$

$$\frac{d^2\varepsilon_1}{dY^2}-M^2\varepsilon_1=0 \quad (9)$$

$$\gamma\frac{d^2\varepsilon_2}{dY^2}-\delta M^2\varepsilon_2=0 \quad (10)$$

$$\left.\begin{array}{l}\varepsilon_1(0)=0;\varepsilon_2(H)=1,\\ \varepsilon_1(1)=\varepsilon_2(1),\dfrac{d\varepsilon_1(1)}{dY}=\gamma\dfrac{d\varepsilon_2(1)}{dY}\end{array}\right\}. \quad (11)$$

$$\frac{\partial V_1}{\partial t}=-P_1+\frac{\partial^2 V_1}{\partial Y^2}-M^2V_1 \quad (12)$$

$$\frac{\partial V_2}{\partial t}=-P_2+\gamma\alpha\frac{\partial^2 V_2}{\partial Y^2}-\alpha\delta M^2V_2 \quad (13)$$

$$V_1(t,0)=0,V_2(t,H)=0,V_1(t,1)=V_2(t,1)$$

$$\frac{dV_1(t,1)}{dY} = \mu_r \frac{dV_2(t,1)}{dY} \tag{14}$$

$$V_i(0,Y) = F_i(Y) - \phi_i(Y) f_1(0) - \varepsilon_i(Y) f_2(0) = F_i^*, i = 1,2 \tag{15}$$

$$P_1 = P + \phi_1 \frac{df_1}{dt} + \varepsilon_1 \frac{df_2}{dt} \tag{16}$$

$$P_2 = P\alpha + \phi_2 \frac{df_1}{dt} + \varepsilon_2 \frac{df_2}{dt} \tag{17}$$

The solutions of equation (6-8) are

$$\phi_1(Y) = A_1 Exp(MY) + A_2 Exp(-MY) \tag{18}$$

$$\phi_2(Y) = A_3 Exp\left(MY\sqrt{\frac{\delta}{\gamma}}\right) + A_4 Exp\left(-MY\sqrt{\frac{\delta}{\gamma}}\right) \tag{19}$$

where $A_1, A_2, A_3,$ and A_4 are found from the solution of the following set of equations:

$$\begin{bmatrix} 1 & 1 & 0 & 0 \\ Exp(M) & Exp(-M) & -Exp\left(M\sqrt{\frac{\delta}{\gamma}}\right) & -Exp\left(M\sqrt{\frac{\delta}{\gamma}}\right) \\ Exp(M) & -Exp(-M) & -\sqrt{\gamma\delta} Exp\left(M\sqrt{\frac{\delta}{\gamma}}\right) & \sqrt{\gamma\delta} Exp\left(-M\sqrt{\frac{\delta}{\gamma}}\right) \\ 0 & 0 & Exp\left(MH\sqrt{\frac{\delta}{\gamma}}\right) & Exp\left(-MH\sqrt{\frac{\delta}{\gamma}}\right) \end{bmatrix} \begin{bmatrix} A_1 \\ A_2 \\ A_3 \\ A_4 \end{bmatrix} = \begin{bmatrix} 1 \\ 0 \\ 0 \\ 0 \end{bmatrix} \tag{20}$$

Also, the solutions of equation (9-11) are

$$\varepsilon_1(Y) = A_5 Exp(MY) + A_6 Exp(-MY) \tag{21}$$

$$\varepsilon_2(Y) = A_7 Exp\left(MY\sqrt{\frac{\delta}{\gamma}}\right) + A_8 Exp\left(-MY\sqrt{\frac{\delta}{\gamma}}\right) \tag{22}$$

where $A_5, A_6, A_7,$ and A_8 are found from the solution of the following set of equations:

$$\begin{bmatrix} 1 & 1 & 0 & 0 \\ Exp(M) & Exp(-M) & -Exp\left(M\sqrt{\frac{\delta}{\gamma}}\right) & -Exp\left(M\sqrt{\frac{\delta}{\gamma}}\right) \\ Exp(M) & -Exp(-M) & -\sqrt{\gamma\delta}\,Exp\left(M\sqrt{\frac{\delta}{\gamma}}\right) & \sqrt{\gamma\delta}\,Exp\left(-M\sqrt{\frac{\delta}{\gamma}}\right) \\ 0 & 0 & Exp\left(MH\sqrt{\frac{\delta}{\gamma}}\right) & Exp\left(-MH\sqrt{\frac{\delta}{\gamma}}\right) \end{bmatrix} \begin{bmatrix} A_5 \\ A_6 \\ A_7 \\ A_8 \end{bmatrix} = \begin{bmatrix} 0 \\ 0 \\ 0 \\ 1 \end{bmatrix} \tag{23}$$

Using Green's function approach, Equation (12-15) assumes the following form:

$$\begin{aligned} V_1(t,Y) = & \int_0^1 G_{11}(Y,t/Y',t')_{t'=0} F_1^*(Y')dY' \\ & - \int_{t'=0}^{t} dt' \int_0^1 G_{11}(Y,t/Y',t')(P_1)dY' \\ & + \int_1^H G_{12}(Y,t/Y',t')_{t'=0} F_2^*(Y')dY' \\ & - \int_{t'=0}^{t} dt' \int_1^H G_{12}(Y,t/Y',t')(P_2)dY' \end{aligned} \tag{24}$$

$$\begin{aligned} V_2(t,Y) = & \int_0^1 G_{21}(Y,t/Y',t')_{t'=0} F_1^*(Y')dY' \\ & - \int_{t''=0'}^{t} dt' \int_0^1 G_{21}(Y,t/Y',t')(P_1)dY' \\ & + \int_1^H G_{22}(Y,t/Y',t')_{t'=0} F_2^*(Y')dY' \\ & - \int_{t''=0}^{t} dt' \int_1^H G_{22}(Y,t/Y',t')(P_2)dY' \end{aligned} \tag{25}$$

where G_{ij} is the appropriate Green's function found from the homogeneous version of the governing equations (12-15) are:

$$G_{ij}(Y,t/Y',t') = \sum_{n=1}^{\infty} \frac{Exp\left(-\beta_n^2(t-t')\right)}{N_n} \eta_i \psi_{in}(Y)\psi_{jn}(Y').$$

where

$$i=1,2 \;\&\; j=1,2, \eta_1 = 1, \eta_2 = \frac{1}{\alpha}. \tag{26}$$

$$N_n = \int_0^1 \psi_{1n}^2(Y')dY' + \left(\frac{1.}{\alpha}\right)\int_1^H \psi_{2n}^2(Y')dY' \tag{27}$$

$$\psi_{1n}(Y) = Sin(C_n Y), \text{ where } C_n^2 = \beta_n^2 - M^2 \tag{28}$$

$$\psi_{2n} = A_{2n} Cos(\lambda_n Y) + B_{2n} Sin(\lambda_n Y), \text{ where } \lambda_n^2 = \left[\frac{\beta_n^2}{\alpha\gamma} - \frac{\delta M^2}{\gamma}\right] \tag{29}$$

The values of constant A_{2n} and B_{2n} are found from the solution of following set of equations

$$\begin{bmatrix} Cos(\lambda_n) & Sin(\lambda_n) \\ -\gamma\lambda_n Sin(\lambda_n) & \gamma\lambda_n Sin(\lambda_n) \end{bmatrix} \begin{bmatrix} A_{2n} \\ B_{2n} \end{bmatrix} = \begin{bmatrix} Sin(C_n) \\ C_n Cos(C_n) \end{bmatrix} \tag{30}$$

The eigen-values λ_n and β_n are found as the roots of

$$A_{2n} Cos(H\lambda_n) + B_{2n} Sin(H\lambda_n) = 0. \tag{31}$$

3. PARTICULAR CASE

The physical situation, in which flow inside the channel is solely caused by uniform pressure gradient, i.e., $F_1(Y) = F_2(Y) = f_1(t) = f_2(t) = 0$, which yields:

$$P_1 = P, P_2 = P\alpha \text{ and } U_i = V_i, i = 1,2.$$

Using these values in equations (22) and (23) the dimensional velocity in both phases are:

$$U_1(Y,t) = -\int_{t'}^{t} dt' \int_0^1 G_{11}(Y,t/Y',t')\psi_{1n}(Y)\psi_{1n}(Y')P dY' - \int_{t'}^{t} dt' \int_1^H G_{12}(Y,t/Y',t')\psi_{1n}(Y)\psi_{2n}(Y')P\alpha dY' \tag{32}$$

$$U_2(Y,t) = -\int_{t'}^{t} dt' \int_0^1 G_{21}(Y,t/Y',t')\psi_{2n}(Y)\psi_{1n}(Y')P dY' - \int_{t'}^{t} dt' \int_1^H G_{22}(Y,t/Y',t')\psi_{2n}(Y)\psi_{2n}(Y')P\alpha dY' \tag{33}$$

$$U_1(Y,t) = -P\sum_{n=1}^{\infty}\frac{[1. - Exp(-\beta_n^2 t)]}{\beta_n^2 N_n C_n}Sin(C_n Y)[1. - Cos(C_n)]$$

$$-P\sum_{n=1}^{\infty}\frac{[1. - Exp(-\beta_n^2 t)]}{\beta_n^2 N_n \lambda_n}Sin(C_n Y)$$

$$[A_{2n}\{Sin(\lambda_n H) - Sin(\lambda_n)\} - B_{2n}\{Cos(\lambda_n H) - Cos(\lambda_n)\}] \quad (34)$$

$$U_2(Y,t) = -P\sum_{n=1}^{\infty}\frac{[1. - Exp(-\beta_n^2 t)][A_{2n}Cos(C_n Y) + B_{2n}Sin(C_n Y)][1. - Cos(C_n)]}{\beta_n^2 N_n C_n}$$

$$-P\sum_{n=1}^{\infty}\frac{[1. - Exp(-\beta_n^2 t)][A_{2n}Cos(C_n Y) + B_{2n}Sin(C_n Y)]}{\beta_n^2 N_n \lambda_n}$$

$$[A_{2n}\{Sin(\lambda_n H) - Sin(\lambda_n)\} - B_{2n}\{Cos(\lambda_n H) - Cos(\lambda_n)\}] \quad (35)$$

$$N_n = \int_0^1 \psi_{1n}^2(Y')dY' + \left(\frac{1.}{\alpha}\right)\int_1^H \psi_{2n}^2(Y')dY'$$

$$= \left[\frac{Y'}{2} - \frac{Sin(2C_n Y')}{4}\right]_0^1 + \left(\frac{A_{2n}^2}{\alpha}\right)\left[\frac{Y'}{2} - \frac{Sin(2\lambda_n Y')}{4\lambda_n}\right]_1^H$$

$$+\left(\frac{B_{2n}^2}{\alpha}\right)\left[\frac{Y'}{2} - \frac{Sin(2\lambda_n Y')}{4\lambda_n}\right]_1^H - \left(\frac{A_{2n}B_{2n}}{\alpha\lambda_n}\right)[Cos(2\lambda_n Y')]_1^H \quad (36)$$

4. Steady State Solution

$$\frac{\partial^2 U_1}{\partial Y^2} - M^2 U_1 = P \quad , \quad (37)$$

$$\gamma\frac{\partial^2 U_2}{\partial Y^2} - M^2 U_2 = P \quad , \quad (38)$$

$$U_1 = C_1 Sinh(MY) - \frac{P}{M^2}[Cosh(MY) - 1.] \quad , \quad (39)$$

$$U_2 = C_3 Cosh\left(MY\sqrt{\frac{\delta}{\gamma}} \right) + C_4 Sinh\left(MY\sqrt{\frac{\delta}{\gamma}} \right) - \frac{P}{M^2\delta}, \quad (40)$$

where C_1, C_3 and C_4 are found from the solution of following set of equations:

$$\begin{bmatrix} 0 & Cosh\left(MH\sqrt{\frac{\delta}{\gamma}} \right) & Sinh\left(MH\sqrt{\frac{\delta}{\gamma}} \right) \\ -Sinh(M) & Cosh\left(M\sqrt{\frac{\delta}{\gamma}} \right) & Sinh\left(M\sqrt{\frac{\delta}{\gamma}} \right) \\ -MCosh(M) & M\sqrt{\gamma\delta}Sinh\left(M\sqrt{\frac{\delta}{\gamma}} \right) & M\sqrt{\gamma\delta}Cosh\left(M\sqrt{\frac{\delta}{\gamma}} \right) \end{bmatrix} \begin{bmatrix} C_1 \\ C_3 \\ C_4 \end{bmatrix} = \begin{bmatrix} Z_1 \\ Z_2 \\ Z_3 \end{bmatrix} \quad (41)$$

where

$$Z_1 = \frac{P}{M^2\delta}$$

$$Z_2 = \frac{P}{M^2}\left[Cosh(M) - 1. \right] + Z_1$$

$$Z_3 = \frac{P}{M} Sinh(M)$$

Concluding Remarks

The novel feature of the work is to present an a closed form solution of time-dependent MHD two phase flows where both phases are electrically conducting and having different physical properties using Green's function approach.

List of Symbols

d = width of the non-conducting fluid
h = total width of channel

H = dimensionless total width of the channel $\left(\frac{h}{d} \right)$

B_0 = magnetic field strength

u_1 = velocity field of conducting fluid in phase I

u_2 = velocity field of conducting fluid in phase II

U_1 = dimensionless velocity field of conducting fluid in phase I $\left(\frac{u_1 d}{\nu_1}\right)$

U_2 =dimensionless velocity field of conducting fluid in phase II $\left(\frac{u_2 d}{\nu_1}\right)$

y =dimensional transverse co-ordinate

Y = dimensionless transverse co-ordinate $\left(\frac{y}{d}\right)$

$\frac{\partial p}{\partial x}$ =dimensional axial pressure gradient

P =dimensionless axial pressure gradient $\left(\frac{\partial p}{\partial x}\left(\frac{d^3}{\rho_1 \nu_1^2}\right)\right)$

M =Hartman number $\left(B_0 d\sqrt{\frac{\sigma}{\mu_1}}\right)$

$f_1^*(\tau)$= velocity of lower bounding wall
$f_2^*(\tau)$= velocity of upper bounding wall
$\overline{F_1}(y)$ = initial velocity in non-conducting fluid
$\overline{F_2}(y)$ = initial velocity in conducting fluid

$f_1(t)$=dimensionless velocity of lower bounding wall $\left(\frac{f_1^*(\tau)d}{\nu_1}\right)$

$f_2(t)$=dimensionless velocity of upper bounding wall $\left(\frac{f_2^*(\tau)d}{\nu_1}\right)$

$F_1(Y)$=dimensionless initial velocity in non-conducting fluid $\left(\frac{\overline{F_1}(y)d}{\nu_1}\right)$

$F_2(Y)$=dimensionless initial velocity in conducting fluid $\left(\frac{\overline{F_2}(y)d}{\nu_1}\right)$

t =dimensionless time $\left(\frac{\tau \nu_1}{d^2}\right)$

Greek Symbols

τ = time
μ =dynamic viscosity

ν = kinematic viscosity

ρ =density of fluid

σ =electrical-conductivity of conducting fluid

γ =ratio of dynamic viscosities $\left(\dfrac{\mu_2}{\mu_1}\right)$

α =ratio of densities of fluids $\left(\dfrac{\rho_1}{\rho_2}\right)$

δ =ratio of electrical conductivities of fluids $\left(\dfrac{\sigma_2}{\sigma_1}\right)$

Subscripts
1= conducting fluid phase I
2= conducting fluid phase II

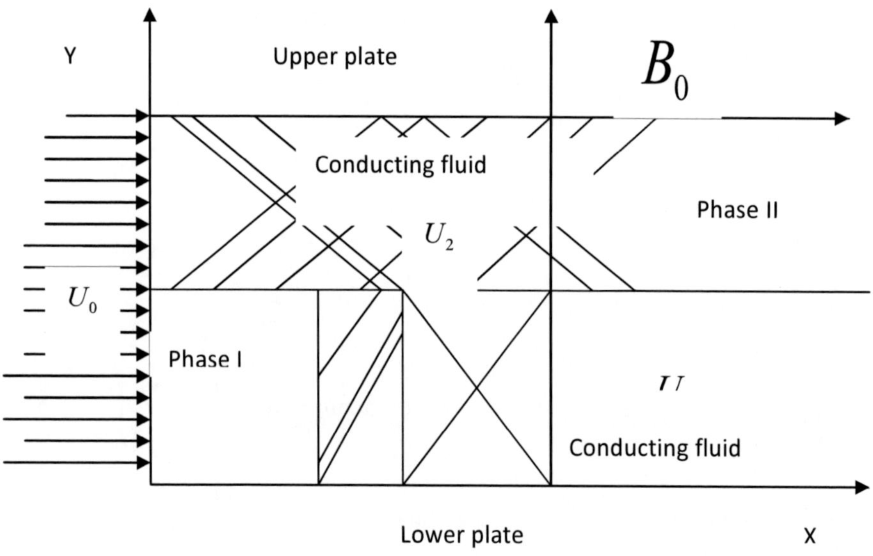

Figure 1. Physical Configuration.

REFERENCES

[1] R.Shail., "On laminar two phase flow in magneto-hydrodynamics" *International Journal of Engineering Sciences*, 11(1973)1103-1108.

[2] J.Lohrasbi and V.Sahai., "Magneto-hydrodynamic heat transfer in two-phase flow between parallel plates"., *Applied Scientific Research,* 45,(1988)pp53-66.

[3] M.S.Malashetty and V. Leela., "Magneto-hydrodynamic heat - transfer in two-fluid flow", *Proc.ASME/AIChE 27th National Heat transfer Conference and Exposition*,28-31 july(1991).

[4] M.S.Malashetty and V. Leela., "Magneto-hydrodynamic heat transfers in two-phase flow", *International Journal of Engineering Sciences,* 30(1992)371-377.

In: New Developments in Magnetohydrodynamics …
Editors: Stephen Talton and Patrick F. Theba

ISBN: 978-1-61942-078-6
2012 Nova Science Publishers, Inc.

Chapter 8

A NOTE ON THERMONUCLEAR HOT BIG BANG WITH REFERENCE TO CONTROLLED FUSION OF THE SUN

S. K. Ghosh[1*] and O. Anwar Bég[2]

[1]Department of Mathematics, Narajole Raj College P.O. – Narajole, Dist. – Midnapore (West), Pin Code – 721 211 West Bengal, INDIA,

[2]Magnetohydrodynamics and Heat Transfer Group, Mechanical Engineering Department, Harmer Building, Sheffield Hallam University, Sheffield, South Yorkshire, S11WB, UK

ABSTRACT

Resonance exhibits a controlled thermonuclear fusion reaction of the Sun in vacuum permeated by a magnetic mirror with a decisive importance to a driving force. An abundance of helium with the Sun is exposed to a nuclear bombardment with ionized hydrogen with reference to a chain reaction so that hydrogen bomb blasting on the surface of the Sun is reliably exhibited. With reference to a driving force it is rigorously stated that a laser radiation communicates ultraviolet radiation and X – ray emission at the resonant level as the magnetic field grows towards the resonant level with reference to a radio frequency accelerator with the Sun.

Keywords: Hot Big Bang, White dwarf, Supernova, binary X-ray, Laser radiation.

ASTROPHYSICAL INTERPRETATIONS WITH TECHNICAL REMARKS

A controlled thermonuclear fusion reaction of the Sun deals with the excitation frequency at the resonant level in the presence of a pressure gradient if a forced oscillation is taken into account. Since the position of the Sun becomes unstable due to a driving force caused by a forced oscillation, the differential rotation of the Sun affects the formation of galaxies. A Hot Big Bang of the universe deals with an irregular change of galaxies due to a controlled thermonuclear fusion reaction of the Sun permeated by a magnetic mirror with reference to E

* Email : g_swapan2002@yahoo.com

$>> mc^2$ where E is the energy , m is the mass and c is the velocity of light. An irregular change of galaxies communicates spiral radio galaxies where many of the giant stars are moving fast and as a result, the collision of the giant stars leads to an explosion to become white dwarf so that the universe becomes cool. It is more convincing to say that the collapse of a giant star leads to a supernova explosion due to thermonuclear fusion reaction of the Sun. Therefore, the collapse of a giant star becomes white dwarf. A thermonuclear Hot Big Bang is an exposure of γ - radiation of neutron stars to raise its temperature as neutron stars expand to generate explosion. It reveals that white dwarf has become a significant effect to generate a resonant line to form a neutron star with ultrarelativistic electrons that raises its temperature and the neutron star grows to form the galaxies to determine the young universe. A radiofrequency accelerator with the Sun breaks the frictional layer at the boundary of the solar corona to initiate thermonuclear fusion reaction in a vacuum in the presence of a magnetic mirror. A gravitational pull is attracted by the Sun and the Earth tends to go slowly and slowly nearer to the Sun so that the stars and planets become irregular. An irregular change of galaxies corresponds to spiral radio galaxies where many of the giant stars move with an irregular space – time interval and some of the giant stars carry high energy radiation and the collision of giant stars leads to a supernova explosion to become a neutron star. Thus an explanation of white dwarf becomes relevant to the case of a neutron star that raises its temperature to form a young star which is compatible with the Sun with a decisive importance to an alternative configuration of the Sun. It seems to be understood that the new formation of the universe would be detected by white dwarf. Although the present universe experiences an abundance of helium with the Sun to a nuclear bombardment with ionized hydrogen due to a chain reaction so that the continuous glowing of ionized hydrogen with irregular fluctuation of electrons becomes relevant to the case of a strong ionizing radiation in the atmosphere. A radiofrequency accelerator with the Sun exerts its influence of a flood of water on the surface of the Moon due to an irregular fluctuation of electrons in the presence of a magnetic mirror with the Sun. This situation reveals that the differential rotation of the Sun affects the Moon and the Earth with reference to a controlled thermonuclear fusion reaction of the Sun to expose artificial rain fall and the strength of the magnetic field is a deterministic realization of artificial rain fall as the magnetic field grows towards the resonant level. A controlled thermonuclear fusion reaction of the Sun in the presence of a magnetic mirror can lead to a microwave background of radiation with reference to $E >> mc^2$ if a magnetic field grows to an appropriate level. A Hot Big Bang of the universe deals with three independent pieces of evidences viz, 1) the expansion of universe, 2) a microwave background of radiation and 3) abundance of Helium. Due to irregular change of galaxies with reference to $E >> mc^2$ it is stated that a microwave background of radiation does exist outer of the solar system. Nevertheless a microwave background of radiation plays an important role of the solar system due to a Hot Big Bang of the universe so that the existence of ultraviolet radiation has become a remarkable evidence of the planet "Mars"(See Ghosh [5]) due to an irregular change of galaxies with reference to $E >>mc^2$. In turn neurocybernetics of the Sun breaks microwave background of radiation and the universe become trapped by X – emission at the resonant level. In taking into account it is rigorously stated that nuclear bombardment with ionized hydrogen appears to be an emergent situation of Nitrogen di – oxide (NO_2) in the atmosphere. Due to abundance of Helium with the Sun, ionization takes place to an account of waste product of Carbon di – oxide (CO_2) in the Earth atmosphere. This falls to a reason of an environmental degradation of the Earth. Although a remarkable attention of the universe have been carried out by Ghosh [1], Ghosh and Pop [2 - 3] and Ghosh [4 – 6] and Ghosh et. al [7] it is necessary to point out some technical remarks on this field. Thermonuclear Hot Big Bang leads to rise in temperature at the polar region with large-scale wave disturbances with reference to the growth of a magnetic field at the resonant level. The expansion of universe experiences a thermonuclear Hot Big Bang to deal with irregular change of galaxies where stars and planets become irregular. Since Interstellar medium is a turbulent model, irregularities of stars lead to go far away from the outer space. The planets move slowly and slowly with irregular space-time interval to go far away from the solar family. The best example of a planet "Pluto" is out of range from the solar system. A serious problem arising out of a thermonuclear Hot Big Bang of the universe that leads to a supernova explosion with the Sun due to the presence of a pressure gradient in a vacuum. A magnetic mirror with the Sun initiates thermonuclear fusion reaction due to an abundance of Helium and an oscillator takes place to produce auto oscillations in the presence of a radiofrequency accelerator with a relay circuit. Nevertheless the magnetic mirror is an exposure of a controlled thermonuclear fusion reaction of the Sun as defined by laser radiation due to a driving force with reference to the growth of magnetic field lines. A controlled thermonuclear fusion reaction of the Sun at the resonant level due to a driving force in the presence of a magnetic mirror to produce multi photon with the inclusion of ultraviolet radiation and binary X-rays are emitted as the air breaks down (Vacuum) to form an ionized plasma. It is observed that a controlled thermonuclear fusion reaction of the Sun leads to an oscillating current flow where radio emission is reliability exhibited so that radio emission is a synchrotron radiation, the emission of ultrarelativistic electrons gyrating in magnetic fields. It is noticed that the Navigation with reference to radar communication system with telecommunication network is the synchronization of atmospheric signal phenomenon via radio emission. Therefore gyroscopic system can lead to a differential rotation of the Sun. A back thermonuclear fusion reaction of the Sun is essentially true for the case to setup a magnetic mirror with the Sun and the behavior is therefore to reinforce in a vacuum (back of the Sun) as some of the planets does exist in a vacuum. In turn, a secret planet does exit in a vacuum with a high frequency radio transmitter. A controlled thermonuclear fusion reaction with the Sun exerts its influence of a neurocybernetics to deal with radio emission in a vacuum. A thermonuclear Hot Big Bang of the universe is trapped by X- ray emission and a laser radiation is confined to a resonant level due to a driving force with a band of ultraviolet radiation and X- ray emission of shorter wave length. An outstanding contribution to the appliance of a magnetic mirror with the Sun exerts its influence of an inclined magnetic field at the resonant level. Ghosh [1], Ghosh and Pop [2 - 3] and Ghosh et. al [7] observed that the oscillatory dynamo action of the universe can lead to a resonant condition $\omega > \frac{1}{2} \cos \omega T\ (16\ K^4 - M^4 \sin^4\theta)^{1/2}$. With reference to an oscillator with the Sun, it is rigorously stated that oscillation builds up with a driving force to an angular frequency of oscillation $\omega T = \pi/2$ and its reflection occurs at resonance as the magnetic field increases in strength abruptly. Therefore, a resonant condition $\omega > \frac{1}{2} \cos \omega T\ (16\ K^4 - M^4 \sin^4\theta)^{1/2}$ is to be satisfied with reference to $\omega T = \pi/2$ where the driving frequency $\omega > 0$. This situation reveals that a magnetic mirror can lead to intense laser radiation and its amplification is very high. It is important to note that a radio frequency accelerator with the Sun communicates laser radiation with a driving force to produce the growth of a magnetic field at the resonant level so that a band of ultraviolet radiation and x – emission of shorter wave length is reliably exhibited. It reveals that a photon light can be transformed into laser pulse radiation permeated by a radiofrequency accelerator with reference to a driving force to take into consideration of laser rainbow with a decisive importance to an oscilloscope as the magnetic field grows towards the resonant level. In response to the above resonant condition it is more convincing to say that a controlled thermonuclear fusion reaction of the Sun leads to a turbulent regime in taking into account of a turbulent dynamo action of the universe with reference to solar and terrestrial context if a magnetic force and Coriolis force are comparable of magnitude. Obviously, a magnetic field grows at an appropriate level. A radio frequency accelerator and controlled fusion become significant to solar and terrestrial context of the universe permeated by a magnetic mirror. Nuclear synthesis deals with ionized hydrogen to transform into abundance of helium with the Sun due to nuclear bombardment with the Sun so that radio emission can be transformed into a relativistic ionized plasma with regard to $E >> mc^2$. This situation reveals that a spiral radio galaxies may find some of the living stars with an alternative configuration of the Sun and white dwarf becomes a new finding of an universe of the outer space. A radio frequency accelerator with the Sun exerts its influence of buoyancy force on the surface of the Sun with an increase in differential rotation of the Sun. The simultaneous action of buoyancy force and Coriolis force plays an important role of an oscillatory dynamo mechanism of the system at the resonant level permeated by a magnetic mirror. Owing to increase in differential rotation of the Sun, a magnetic mirror takes place of cyclonic turbulence with an abrupt increase in magnetic field and the magnetic field changes its direction abruptly in the central region. A controlled thermonuclear fusion reaction of the Sun at the resonant level in the presence of a magnetic mirror communicates helical plasma waves in the universe propelled by a Hot Big Bang. The simultaneous affects on Coriolis force and buoyancy force plays a significant role on helicity generation of waves. The best example of a turbulent dynamo model of the universe to deal with M – theory is the representation of a helical plasma waves. A radio signal from the Sun to exert its influence of helical plasma waves leading to an application of telecommunication such as navigation as well as Ocean science around the polar region is a new expedition of plasma science. A controlled thermonuclear fusion reaction of the Sun is an expedition of outer space to deal with species at 0 to – 273°C. A microwave background of radiation at the resonant level does exit at the planet "Mars" to communicate with species. In this context species grows with ultraviolet radiation to form a living universe with neurocybernetics with reference to a magnetic mirror in taking into account of the origin of species "dinosaur". Eventually a controlled thermonuclear fusion reaction of the Sun permeated by a magnetic mirror is a realistic feature of the universe to find some of the planets into the spiral radio galaxies and white dwarf plasma becomes a new finding of outer universe to enter into a new space age at 0° C to – 273° C. A planet in the outer universe may come into true which is nearer to Sun (Ice globe) at - 273° C with a decisive importance to the back of the Sun (Vacuum).

REFERENCES

[1] Ghosh, S. K., A note on unsteady hydromagnetic flow in a rotating channel permeated by an inclined magnetic field in the presence of an oscillator, *Czechoslovak Journal of Physics*,.51, No. 8, 799-804 (2001).

[2] Ghosh, S. K. and Pop, I., Comments on MHD plasma flow in a rotating environment in the presence of an inclined magnetic field, *International Journal of Applied Mechanics and Engineering,* 9, No. 4, pp. 821 – 826 (2004).

[3] Ghosh S. K. and Pop, I., An analytical approach to MHD plasma behavior of a rotating environment in the presence of an inclined magnetic field as compared to excitation frequency, *International Journal of Applied Mechanics and Engineering,* 11, No. 4, 845- 856 (2006)

[4] Ghosh, S. K., Report on plasma behavior over the Interstellar medium, *International Journal of Applied Mechanics and Engineering*, 11, No. 4, pp. 981-983 (2006).

[5] Ghosh, S. K., Exploitation of the universe with regard to controlled thermonuclear fusion reaction of the Sun. – *International Journal of Applied Mechanics and Engineering* , 12 , No. 2, pp. 571- 573 (2007).

[6] Ghosh, S. K., Transmission of radio wave with regard to controlled thermonuclear fusion reaction of the Sun. - *Adv. Studies Theor. Phys.* , 1, No. 4 , pp. 169 – 172. (2007).

[7] Ghosh, S. K., Beg, O.A., Zueco, J. and Prasad, V. R., *Transient hydromagnetic flow in a rotating channel permeated by an inclined magnetic field with magnetic induction and Maxwell displacement current effects, Zeitschrift angewandte Mathematik und Physik* (ZAMP), Vol. 61, No. 1, pp. 147 – 169 , 2010.

In: New Developments in Magnetohydrodynamics ... ISBN: 978-1-61942-078-6
Editors: Stephen Talton and Patrick F. Theba

Chapter 9

LARGE HYDROMAGNETIC MIXED CONVECTION FLOW OF VISCOUS INCOMPRESSIBLE FLUID PAST A MAGNETIZED VERTICAL POROUS PLATE

Muhammad Ashraf and Md. Anwar Hossain*
Department of Mathematics, University of Dhaka, Dhaka, Bangladesh.
Department of Mathematics, COMSATS Institute of Information Technology
Islamabad, Pakistan

Abstract

This paper is concerned with the steady mixed convection boundary layer flow of viscous incompressible electrically conducting fluid past a semi infinite magnetized vertical porous plate with uniform transpiration and variable transverse magnetic field along the surface. The equations governing the flow, magnetic and temperature field are reduced to dimensionless convenient form using the free variable transformations and solved numerically by using finite difference method. Effects of the physical parameters, such as, Pr, S, and λ on the numerical values of skin friction coefficient, $Re_x^{1/2}Cf_x$, heat transfer coefficient, $Re_x^{-1/2}Nu_x$, and coefficient of magnetic intensity, $Re_x^{1/2}Mg_x$ against ξ, the local transpiration parameter are presented in tabular form and the velocity profile, temperature profile and the profile of transverse component of magnetic field are shown graphically. Asymptotic solutions are also obtained for smaller and larger values of ξ and compared with the entire values of ξ.

Keywords: Hydromagnetic, mixed convection, magnetized porous plate, magnetic intensity, heat transfer

*E-mail: anwar@univdhaka.edu

1. Nomenclature

Cf_x	Skin friction
f	Transformed stream function
g	Acceleration due to gravity, $m.s^{-2}$
Gr_x	Local Grashof number
H_x	Magnetic field along the surface
H_y	Magnetic field normal to the surface
H_0	Reference magnetic field velocity
H_∞	Free stream magnetic field
Mg_x	Magnetic intensity at the surface
Nu_x	Local Nusselt number
Pm	Magnetic Prandtl number
Re_x	Local Reynolds number
S	magnetic field parameter
T_w	Wall temperature, K
T_∞	Ambient fluid temperature, K
u	Dimensional axial velocity, $m.s^{-1}$
U_0	Reference velocity, $m.s^{-1}$
U_∞	Free stream velocity, $m.s^{-1}$
v	Dimensional normal velocity, $m.s^{-1}$
V_0	Surface mass flux
x	Axial distance, m
y	Normal distance, m

Greek letters

α	Thermal diffusivity, $m^2.s^{-1}$
β	Volumetric expansion coefficient
γ	Magnetic diffusion
θ	Dimensionless temperature function
λ	Mixed convection parameter
μ	Dynamical viscosity, $Kg.m^{-1}.s^{-1}$
ν	Kinematic viscosity, $m^2.s^{-1}$
ξ	Transpiration parameter
ψ	Fluid Stream function, $m^2.s^{-1}$
ϕ	magnetic potential function
ρ	Density of the fluid, $Kg.m^{-3}$
σ	Electrical conductivity, $S(siemens).m^{-1}$
η	Similarity transformation
$\bar{\mu}$	Magnetic permibility

Subscripts

w	Wall condition
∞	Ambient condition

2. Introduction

The magnetohydrodynamics of an electrically conducting fluid is encountered in many problems in geophysics, astrophysics, engineering applications and other industrial areas. One physical interest in this flow lies in the possibility of using such a field to shield a body from excessive heating. Hydromagnetic mixed convection flow have a great significance for the applications in the stellar atmosphere, planetary magnetospheres and aeronautics. Magnetohydrodynamics principles also being applied in designing heat exchangers, pumps, space vehicle propulsion, thermal protection, and also in nuclear generating systems. In many metallurgical processes involve the cooling of continuous strips or filaments by drawing them through an electrically conducting fluid subject to a magnetic field. With this understanding, Greenspan and Carrier [1] was the first who investigated the flow of viscous, incompressible and electrically conducting fluid in the presence of a symmetrically oriented semi-infinite flat plate in which magnetic field assumed to be coincident with the ambient fluid velocity field. In this investigation fourier transformation together with asymptotic analysis had been incorporated and found that the velocity gradient at the plate approaches zero due to increase in the applied magnetic field intensity. Further contributions to the problem was added by Davies [2] and [3] considering the fact that the flow is opposed by magnetodynamic pressure gradient along an non-magnetized plate and concluded that for the magnetic field parameter (or Chandrashekhar number) $S \simeq 1$ the drag coefficient vanishes. Gribben [4] then considered an axisymmetric magnetohydrodynamic flow of an incompressible, viscous, electrically conducting fluid near a stagnation point considering that the magnetic field lines are circles and parallel to the surface. Later, Gribben [5] who investigated the magnetohydrodynamic boundary layer in steady incompressible flow under the influence of an external magnetic dynamic pressure gradient using the asymptotic analysis and found that the skin friction decreases with the increase of magnetic field. The boundary layer flow and heat transfer of hydromagnetic flow of viscous incompressible fluid flows past an electrically insulated semi-infinite flat plate in the presence of a uniform magnetic field parallel to the plate has been investigated by Ramamoorthy [6] numerically and found that the presence of the magnetic field increases both the momentum and thermal boundary layer thicknesses. On the other hand Tan and Wang [7] studied the effect of applied magnetic field on temperature distribution as well as on the recovery temperature due to the flow of a viscous incompressible electrically conducting fluid past a solid plane surface subject to uniform heat flux. They concluded that the values of recovery factor decreases with the increase of both magnetic field parameter S and magnetic Prandtl number Pm. Hildyard [8] found that the magnetic-field boundary condition used by Gribben was inappropriate and hence making the necessary correction obtained the appropriate asymptotic solutions for large and small values of the magnetic Prandtl number, Pm. Later, Chawla [9] studied the effect of free stream fluctuations on the flow over a semi-infinite plate, with an aligned magnetic field, using von Kármán-Pohlhausen technique and solution for low and high frequency ranges are developed. But, Ingham [10] studied the boundary layer flow on a semi-infinite flat plate placed at zero incidence to a uniform stream of electrically conducting gas with an aligned magnetic field at large distances from the plate. In this analysis the author observed that increasing magnetic field for a given Mach number, or decreasing the Mach number for a given magnetic field thickens the momentum and thermal boundary layer.

In all the above investigations, the surface along which the flow of the fluid were considered as non-magnetized. In recent technological development it is necessary to distorted the attention towards magnetized surface, Glauert [11], first, studied the magnetohydrodynamic boundary layer in uniform flow past a magnetized plate for the small and large values of magnetic Prandtl number, Pm. The observation from this investigation, shows that the velocity and magnetic fields are valid for small value of magnetic field parameter S and for both smaller or larger value of magnetic Prandtl number Pm.

Chawla [12] studied the magnetohydrodynamics boundary layer in uniform flow past a semi-infinite magnetize plate, and a magnetic field fluctuating about a non zero mean in the stream direction, is applied to the plate. He comments that in order to create a fluctuating magnetic field, one needs to join the plate in the manner of Glauert[11] with d.c and a.c generators placed in series. However, Chawla assuming the amplitude of the oscillating transverse magnetic field is much smaller than the uniform magnetic field at the surface. He also considered the basic steady flow using Karman-Pohlhausen technique, and obtained approximate solutions to both steady and oscillating part of the flow.

Boundary layer flows are often controlled by injecting or withdrawing fluid through a porous bounding heated surface. This can lead to enhanced heating or cooling of the system and can help to delay the transition from laminar to turbulent. In the field of aeronautical engineering; it is often necessary to prevent separation in order to reduce drag and to attain high-lift as well as in controlling the boundary-layer thickness and the rate of heat transfer. Steady two-dimensional laminar forced flow and heat transfer from a porous wedge was considered in great detail by Lin *et al.* [13]. Later, Koh and Hartnett [14] studied the incompressible laminar flow over a porous wedge with suction and a variable wall temperature. They obtained similarity solutions for wall temperature and the suction rate variations, which are proportional to a power of x. From the practical point of view, a uniform transpiration velocity may be more easily realized than an $x^{(1-m)/2}$ distribution.

The heat transfer in a laminar boundary layer flow of viscous, electrically conducting and incompressible fluid past a semi infinite porous plate has been investigated subject to a large magnetic number and small transpiration Hossain et al. [15]. Magnetohydrodynamic laminar boundary-layer flow over a wedge with suction or injection (suction/blowing rate is proportional to $x^{(1-m)/2}$ has been investigated by Kafoussias and Nanousis [16] numerically for moderate value of Hartman number. Recently, hydromagnetic flow of viscous incompressible fluid in the presence of adverse magnetohydrodynamic pressure gradient with constant injection or suction across the permeable wedge surface has been discussed by Mahmood *et al.* [17]. They find the solutions of the transformed boundary layer equations employing the local non-similarity method, implicit finite difference method and asymptotic series solution for all, and for both small and large values of the local transpiration parameter ξ.

In the light of above literature survey shows no existence of any study on the boundary layer flow of electrically conducting fluid under both magnetic and buoyancy force along a magnetized plate. Hence in the present article the problem investigated is the hydromagnetic mixed convection laminar boundary layer flow of a viscous, incompressible and electrically conducting fluid along a magnetized permeable surface with a variable magnetic field applied in stream direction at the surface. The boundary layer equations for the momentum, energy and magnetic field are reduced to convenient form for integration using appropriate transformations. The solutions of the transformed boundary layer equations are then simulated employing two methods, namely, (i) finite difference method and the (ii) asymptotic series solution for small and large value of local transpiration parameter $\xi = (V_0 x/\nu)/{Re_x}^{1/2}$ that depends on the surface mass-flux, V_0, as well as the distance x measured from the leading edge of the plate. The pertinent physical parameters that dominate the flow and other physical quantities, such as the local skin-friction Cf_x, rate of heat transfer, Nu_x and the magnetic intensity Mg_x at the surface are the magnetic field parameter, S, and the Prandtl number, Pr and the magnetic Prandtl number Pm and mixed convection parameter λ.

3. Formulation of the Mathematical Model

We consider a steady two-dimensional magnetohydrodynamic mixed convection flow of an electrically conducting, viscous and incompressible fluid past a uniformly heated vertical porous plate. The x-axis is taken along the surface and y-axis is normal to it. A schematic diagram illustrating the flow domain and the coordinate system is shown in Fig. 1. In Fig.1 δ_M and δ_T stands for momentum and thermal boundary layer thicknesses. According to the assumptions, following Davies [2], Tan and Wang [7], the two-dimensional boundary layer equations for the flow, are given as follows:

$$\frac{\partial u}{\partial x} + \frac{\partial v}{\partial y} = 0 \tag{1}$$

$$u\frac{\partial u}{\partial x} + v\frac{\partial v}{\partial y} = \upsilon\frac{\partial^2 u}{\partial y^2} + \frac{\bar{\mu}}{\rho}\left(H_x\frac{\partial H_x}{\partial x} + H_y\frac{\partial H_x}{\partial y}\right) + g\beta(T - T_\infty) \tag{2}$$

$$\frac{\partial H_x}{\partial x} + \frac{\partial H_y}{\partial y} = 0 \tag{3}$$

$$-\gamma\frac{\partial H_x}{\partial y} = (uH_y - vH_x) \tag{4}$$

$$u\frac{\partial T}{\partial x} + V\frac{\partial T}{\partial y} = \alpha\frac{\partial^2 T}{\partial y^2} \tag{5}$$

where u and v are fluid velocity components in x and y-direction respectively, H_x and H_y are the x and y-components of magnetic field, α, $\bar{\mu}$ ρ, ν and γ are the thermal diffusion, magnetic permeability, density, kinematic coefficient of viscosity and magnetic diffusivity of the medium. The solution of the above equations should satisfy the following boundary conditions

$$\begin{aligned} &u(x,0) = 0,\ v(x,0) = \pm V_0,\ H_x(x,0) = H_w(x),\ T(x,0) = T_w \\ &u(x,\infty) = U_\infty(x),\ H_x(x,\infty) = 0,\ T(x,\infty) = T_\infty \end{aligned} \tag{6}$$

The nonlinearity of the momentum, hydromagnetic and energy equation makes it difficult to obtain a closed mathematical solution to the problem. However, by introducing the following non-dimensional dependent and independent variables we have,

$$u = U_0\bar{u},\ v = \frac{v}{L}Re_L^{1/2}\bar{v},\ \ \bar{x} = \frac{x}{L},\ \ \bar{y} = \frac{y}{L}R_{eL}^{\frac{1}{2}}y,\ \ H_x = H_0\bar{H}_x,$$
$$H_y = \frac{H_0}{Re_L^{1/2}}\bar{H}_y,\ \ \theta = \frac{T - T_\infty}{\Delta T} \tag{7}$$

where $\triangle T$ is the temperature difference. By using expression (7) in equations (1)-(6), we have

$$\frac{\partial \bar{u}}{\partial \bar{x}} + \frac{\partial \bar{v}}{\partial \bar{y}} = 0 \tag{8}$$

$$\bar{u}\frac{\partial \bar{u}}{\partial \bar{x}} + \bar{v}\frac{\partial \bar{v}}{\partial \bar{y}} = \frac{\partial^2 \bar{u}}{\partial \bar{y}^2} + S\left(\bar{H}_x\frac{\partial \bar{H}_x}{\partial \bar{x}} + \bar{H}_y\frac{\partial \bar{H}_x}{\partial \bar{y}}\right) + \lambda\bar{\theta} \tag{9}$$

$$\frac{\partial \bar{H}_x}{\partial \bar{x}} + \frac{\partial \bar{H}_y}{\partial \bar{y}} = 0 \tag{10}$$

$$-\frac{1}{Pm}\frac{\partial \bar{H}_x}{\partial \bar{y}} = (\bar{u}\bar{H}_y - \bar{v}\bar{H}_x) \tag{11}$$

$$\bar{u}\frac{\partial \bar{\theta}}{\partial \bar{x}} + \bar{V}\frac{\partial \bar{\theta}}{\partial \bar{y}} = \frac{1}{Pr}\frac{\partial^2 \bar{\theta}}{\partial \bar{y}^2} \tag{12}$$

where

$$Re_L = \frac{U_0 L}{\nu},\ \ Gr_L = \frac{g\beta\Delta T L^3}{\nu^2},\ \ \mathrm{Pr} = \frac{\nu}{\alpha}$$
$$\lambda = \frac{Gr_L}{Re_L^2},\ \frac{1}{\gamma} = \sigma\bar{\mu},\ \ \alpha = \frac{K}{\rho C_p},\ \ S = \frac{\bar{\mu}H_0^2}{\rho U_0^2} \tag{13}$$

where Re_L is the Reynolds number, Gr_L the Grashof number, L the reference length, λ is the mixed convection parameter, Pr the Prandtl number and S the magnetic field parameter (also known as Chandrasekhar number), γ the magnetic diffusion and α is the thermal diffusion.

4. Method of Solutions

To get the set of equations in convenient form for integration, we will introduce the following one parameter group of transformation for the dependent and independent variables:

$$\bar{u} = U(\xi, Y),\ \bar{v} = x^{\frac{1}{2}}(V - \xi),\ \varphi = x^{\frac{1}{2}}\phi,\ \bar{\theta} = x^{-1}\theta(\xi, Y)$$
$$Y = x^{-\frac{1}{2}}\bar{y},\ \xi = Sx^{\frac{1}{2}},\ \bar{H}_x = \frac{\partial \varphi}{\partial y},\ \bar{H}_y = -\frac{\partial \varphi}{\partial x} \tag{14}$$

where φ is the potential function that satisfies the equation (10). By using this group of transformations, which satisfies equation of continuity and by using in (8)-(12): we have set of equations:

$$\frac{1}{2}\xi\frac{\partial U}{\partial \xi} - \frac{1}{2}Y\frac{\partial U}{\partial Y} + \frac{\partial V}{\partial Y} = 0 \tag{15}$$

$$\frac{\partial^2 U}{\partial Y^2}+\lambda\theta = \frac{1}{2}\xi U\frac{\partial U}{\partial \xi}+\left(V-\xi-\frac{1}{2}YU\right)\frac{\partial U}{\partial Y}-S\left[-\frac{1}{2}\phi\frac{\partial^2\phi}{\partial Y^2}+\frac{1}{2}\xi\left(\frac{\partial\phi}{\partial Y}\frac{\partial^2\phi}{\partial\xi\partial Y}-\frac{\partial^2\phi}{\partial Y^2}\frac{\partial\phi}{\partial\xi}\right)\right] \tag{16}$$

$$\frac{1}{Pm}\frac{\partial^2\phi}{\partial Y^2}=\frac{1}{2}U\phi+\frac{1}{2}\xi U\frac{\partial\phi}{\partial\xi}+\left(V-\xi-\frac{1}{2}YU\right)\frac{\partial\phi}{\partial Y} \tag{17}$$

$$\frac{1}{Pr}\frac{\partial^2\theta}{\partial Y^2}=\frac{1}{2}\xi U\frac{\partial\theta}{\partial\xi}+\left(V-\xi-\frac{1}{2}YU\right)\frac{\partial\theta}{\partial Y}-U\theta \tag{18}$$

The appropriate boundary conditions satisfied by the above system of equations are

$$\begin{aligned} &U(\xi,0)=V(\xi,0)=0,\ \phi'(\xi,0)=1,\ \theta(\xi,0)=1\\ &U(\xi,\infty)=1,\ \phi'(\xi,\infty)=0,\ \theta(\xi,\infty)=0 \end{aligned} \tag{19}$$

In order to find the numerical solution we have discretised the expressions (15)-(18) with boundary conditions (19) by using finite difference method, using backward difference for x-direction and central difference for y-direction out of which we get the following system of tri-diagonal algebraic equations.

$$A_1U_{i+1,j}+B_1U_{i-1,j}+C_1U_{i-1,j}=D_1 \tag{20}$$

where, for momentum equation

$$A_1=1+\frac{1}{2}(V_{i,j}-\xi_i-Y_jU_{i,j})\triangle Y \tag{21}$$

$$B_1=-2+\frac{1}{2}\frac{\xi_i}{\triangle\xi}U_{i,j}\triangle Y^2 \tag{22}$$

$$C_1=1-\frac{1}{2}(V_{i,j}-\xi_i-Y_jU_{i,j})\triangle Y \tag{23}$$

$$\begin{aligned} D_1=&\left[\frac{S}{2\triangle\xi}(H_1)_{i,j}((H_1)_{i,j}-(H_1)_{i,j-1})-\frac{1}{2}\frac{\xi_i}{\triangle\xi}U_{i,j}U_{i,j-1}\right]\triangle Y^2\\ &-\left[\frac{S}{4}((H_1)_{i+1,j}-(H_1)_{i-1,j})+\frac{S}{2}((H_1)_{i,j}-(H_1)_{i-1,j})(H_2)_{i,j}\right]\triangle Y \end{aligned} \tag{24}$$

where $(H_1)_{i,j}=(\frac{\partial\phi}{\partial y})_{i,j}$ and $(H_2)_{i,j}=(\frac{\partial\phi}{\partial\xi})_{i,j}$. Similarly for magnetic potential equation we have

$$A_2U_{i+1,j}+B_2U_{i-1,j}+C_2U_{i-1,j}=D_2 \tag{25}$$

$$A_2=\frac{1}{Pm}+\frac{1}{2}(V_{i,j}-\xi_i-Y_jU_{i,j})\triangle Y \tag{26}$$

$$B_2=-\frac{2}{Pm}-\frac{1}{2}(1+\frac{\xi_i}{\triangle\xi})U_{i,j}\triangle Y^2 \tag{27}$$

$$C_2=\frac{1}{Pm}-\frac{1}{2}(V_{i,j}-\xi_i-\frac{1}{2}Y_jU_{i,j})\triangle Y \tag{28}$$

$$D_2=-\frac{1}{2}\frac{\xi_i}{\triangle\xi}U_{i,j}\phi_{i,j-1}\triangle Y^2 \tag{29}$$

and for the energy equation

$$A_3U_{i+1,j} + B_3U_{i-1,j} + C_3U_{i-1,j} = D_3 \tag{30}$$

where

$$A_3 = \frac{1}{Pr} + \frac{1}{2}(V_{i,j} - \xi_i - \frac{1}{2}Y_jU_{i,j})\triangle Y \tag{31}$$

$$B_3 = -\frac{2}{Pr} - (1 - \frac{1}{2}\frac{\xi_i}{\triangle\xi})U_{i,j}\triangle Y^2 \tag{32}$$

$$C_3 = \frac{1}{Pr} - \frac{1}{2}(V_{i,j} - \xi_i - \frac{1}{2}Y_jU_{i,j})\triangle Y \tag{33}$$

$$D_3 = -\frac{1}{2}\frac{\xi_i}{\triangle\xi}U_{i,j}\theta_{i,j-1}\triangle Y^2 \tag{34}$$

velocity can be calculated directly using equation of continuity (15) as shown below:

$$V_{i,j} = V_{i-1,j} - \frac{1}{2}\left(\xi\frac{\Delta y}{\Delta\xi} - Y_j\right)U_{i,j} + \frac{1}{2}\xi\frac{\Delta Y}{\Delta\xi}U_{i,j-1} - \frac{1}{2}Y_jU_{i-1,j} \tag{35}$$

where i and j denote the grid points along the X and Y directions respectively. These tri-diagonal equations are then solved by Gaussian elimination technique. The computation is started at $\xi = 0.0$, and then marches downstream implicitly.

The computation is started at $\xi = 0$, and then marches downstream implicitly. The ordinary differential equations governing the upstream condition at $\xi = 0$ can be obtained by taking the limit of Eqs. (15)(18) as ξ approaches zero. The associated boundary conditions are Eqs. (19) with $\xi = 0$. This is, in fact, the similarity solution of a hydromagnetic forced-convection boundary layer flow of viscous incompressible fluid past a magnetized non-permeable vertical flat plate. The Y and ξ grids are set at 0.01 and 0.025, and the maximum value of Y is 80 in the following computations after several test runs for convergence.

From the above solutions we get the numerical values of U, V, ϕ and θ and their derivatives, and hence readily we obtain the values of skin-friction, Cf_x, Nusselt number, Nu_x, and magnetic intensity, Mg_x, from the following relations:

$$Re_x^{1/2}Cf_x = U'(\xi,0),\ Re_x^{1/2}Mg_x = -\phi(\xi,0),\ Re_x^{-1/2}Nu_x = -\theta'(\xi,0) \tag{36}$$

Here, the prime denotes differentiation with respect to Y.

5. Results and Discussions

In present investigation we the mixed convection flow of a viscous incompressible and electrically conducting fluid past a magnetized vertical but uniformly heated permeable plate. The equations that govern the flow are reduced to convenient form as given in the form given in Eqs. (15)-(18) along with the boundary conditions (19). Solutions of these equations are obtained using the finite difference method as discussed in the preceding section for a wide ranged values of physical parameters, magnetic parameter, S, Prandtl number Pr that are valid for liquid metals, and the magnetic Prandtl number, Pm, against ξ the locally distributed transpiration (also known as suction) parameter. Below we shall discusse the effects of the aforementioned physical parameters of the profiles of velocity, $U(\eta, \xi)$, transverse component of magnetic field, $\phi'(\eta, \xi)$, and that of temperature, $\theta(\eta, \xi)$ as on the local skin-friction coefficient $Re_x^{1/2}Cf_x$, the coefficient of surface magnetic intensity $Re_x^{1/2}Mg_x$ and the rate of heat transfer $Re_x^{-1/2}Nu_x$ from the surface of the plate.

5.1. Effect of Different Physical Parameters on Skin Friction, Magnetic Intensity and Rate of Heat Transfer

As a first attempt, Tables 1, 2 and 3 exhibiting the numerical values of coefficients of skin friction $Re_x^{1/2}Cf_x$, rate of heat transfer $Re_x^{-1/2}Nu_x$ and magnetic intensity $Re_x^{1/2}Mg_x$ for values magnetic field parameter $S = 0.0, 0.8$ against ξ taking value of buoyancy force parameter $\lambda = 0.5$, magnetic Prandtl number $Pm = 0.1$ and Prandtl number Pr=0.1. From these tables it can easily be seen that an increase in magnetic field parameter S leads to decrease both the coefficients of local skin friction and rate of heat transfer; whereas, this leads to increase in the magnetic intensity. This phenomenon can easily be understood from the fact that when magnetic field parameter S increases, the magnetic energy in the uniform stream rises and the thickness of the momentum boundary layer decreases. Therefore the coefficients of skin friction and heat transfer decreases and magnetic intensity at the surface increases due to increasing magnetic field parameter.

Numerical values of the coefficients of skin friction, heat transfer increases and magnetic intensity for values of buoyancy force parameter $\lambda = 0.0$, 2.5, 5.0, 7.5 and 10 against ξ are shown graphically in figures 2(a), 2(b) and 2c for Pm=0.5, Pr=0.1, and S=0.2. From these figures one can observe increase of the coefficients of skin friction and heat transfer increases and decrease in magnetic intensity at the surface owing to increase in the buoyancy force parameter, λ. It is expected that the buoyancy force enhance the velocity field in the boundary layer region. It should further mentioned that increase in λ, i.e., the buoyancy force acts like pressure gradient and thus increase the the fluid motion, hence both the coefficients of skin friction and heat transfer deserve increase; on the other hand the magnetic intensity should decrease along the streamwise distance ξ.

Figures 3(a), 3(b) and 3(c) depict the numerical value of the coefficients of skin friction, rate of heat transfer and magnetic intensity at the surface for different values of Prandtl number Pr (=0.01, 0.1, 0.71, 7.0 and 10) while $\lambda = 1.0$, $S = 0.2$ and $Pm = 0.1$. In these figures, it is observed that owing to increase of Prandtl number Pr the coefficient of skin friction decreases; whereas coefficients of heat transfer and magnetic intensity increase. It is well know from the earlier investigators that the increase of Pr leads to increase the kinematic viscosity (which ratio of dynamic viscosity to density of the fluid) of the fluid and decrease the thermal diffusion which causes the increase in momentum boundary layer thickness and thins the thermal boundary layer. So, these factors are responsible for the aforementioned phenomena.

Finally, the effect of different value of the magnetic Prandtl number Pm (=1.0, 10.0, 100.0) on the coefficients of skin friction, rate of heat transfer and magnetic intensity at the surface against ξ for Pr=0.1, S=2.0, and λ=1.0. is shown throught figures 4(a), 4(b) and

4(c), respectively. From these figures it can be seen that there is increase in the coefficients of skin friction and heat transfer and decrease in the coefficient of magnetic intensity at the surface when value magnetic Prandtl number Pm increases It can further be seen that increase in coefficients of skin friction and heat transfer is not very remarkable even for larger value of magnetic Prandtl Pm; very remarkable. but it dominates on the magnetic intensity. The reason is that increase of magnetic Prandtl number Pm decreases the magnetic diffusion γ and hence the momentum and thermal boundary layer thicknesses become thinner due to which the coefficients of skin friction and heat transfer increases and magnetic intensity at the surface decreases considerably.

5.2. Effect of Different Physical Parameters on Velocity, Temperature and Magnetic Profiles

Now we will discuss the effects of different physical parameters on the velocity profile $U(\xi, Y)$, temperature profile $\theta(\xi, Y)$, and the profile of transverse component of magnetic field, $\phi'(\xi, Y)$, against similarity variable Y at $\xi = 0.5$. To begin with, numerical values of $U(\xi, Y)$, $\theta(\xi, Y)$ and $\phi'(\xi, Y)$, respectively, are shown in figures 5(a), 5(b) and 5(c) against$Y$ for $\lambda = 0.0, 2.5, 5.0, 7.5, 10.0$ and for two values of S equal to 0.0 and 0.8 while $Pm = 0.1$ and Pr=0.1. In these figures dotted lines and solid lines are, respectively, due to S equal to 0.0 (pure fluid) and 0.8 (presence of magnetic field), respectively. From these figures it is observe that the increase in λ leads to increase in the velocity profile and decrease in the temperature profile as well as the profile of transverse component of magnetic field for both the case when magnetic field is absent ($S = 0$) or present ($S = 0.8$) in the flow field. In this figures, comparison between the broken and solid curve shows that the presence of magnetic field in the flow field of the fluid the profiles reduces the numerical values of $\theta(\xi, Y)$ and $\phi'(\xi, Y)$ all through along the surface of the plate. And the difference between the profiles due to ($S = 0$) and ($S = 0.8$) also decreases with increasing value of (λ). Finally, from Fig. 5(a), one can see that there the velocity profile overshoots the edge value and reaches to some maximum value near the surface of the plate in the both the situations when magnetic field in present or absent.

Finally, figures 6(a), 6(b) and 6(c) depict the values of velocity, $U(\xi, Y)$, temperature, $\theta(\xi, Y)$ and the transverse component of magnetic field, $\phi'(\xi, Y)$, respectively, against Y for different values transpiration parameter ξ=0.1, 1.0, 2.0, 4.0, 8.0, and 10.0 while S= 0.4, λ =1.0 and for Pr=0.1, Pm=0. From these figures that what we see that increase in local transpiration parameter ξ decreases the values of velocity, $U(\xi, Y)$, temperature, $\theta(\xi, Y)$ and the transverse component of magnetic field, $\phi'(\xi, Y)$ which ultimately reduces the boundary layer thicknesses of momentum, thermal and of magnetic intensity.

5.3. Asymptotic Solutions for Small and Large ξ

Now we are heading in finding the solution of the present problem for small and large value of transpiration parameter ξ. To do this we first reduce the equations (1)-(6) to convenient form by introducing the following following transformations:

$$\begin{aligned}
\psi &= x^{\frac{1}{2}}\left[f(\eta,\xi)+\xi\right]\\
\varphi &= x^{-\frac{1}{2}}\phi(\eta,\xi),\ \bar{\theta}=x^{-1}\theta(\eta,\xi)\\
\eta &= x^{-\frac{1}{2}},\ \xi = sx^{\frac{1}{2}}
\end{aligned} \tag{37}$$

where, η is the similarity variable, ξ be the local transpiration parameter and ψ, ϕ are the function which satisfy the equations of conservation of mass and magnetic field such that:

$$u=\frac{\partial\psi}{\partial y},\ v=-\frac{\partial\psi}{\partial x} \tag{38}$$

For withdrawal of fluid $\xi > 0$ whereas for blowing of fluid through the surface of the plate $\xi < 0$. Throughout the present computations, value of ξ has been considered positive.

By using (37) and (38) in (1)-(6), we will obtain the following dimensionless local non-similarity equations:

$$f'''+\frac{1}{2}(ff''-S\phi\phi'')+\xi U''+\lambda\theta=\frac{1}{2}\xi\left[f'\frac{\partial f'}{\partial\xi}-f''\frac{\partial f}{\partial\xi}-S\left(\phi'\frac{\partial\phi'}{\partial\xi}-\phi''\frac{\partial\phi}{\partial\xi}\right)\right] \tag{39}$$

$$\frac{1}{Pm}\phi''+\frac{1}{2}(f\phi'-f'\phi)+\xi\phi'=\frac{1}{2}\xi\left[f'\frac{\partial\phi}{\partial\xi}-\phi'\frac{\partial f}{\partial\xi}\right] \tag{40}$$

$$\frac{1}{Pr}\theta''+\frac{1}{2}f\theta'+f'\theta+\xi\theta'=\frac{1}{2}\xi\left[f'\frac{\partial\theta}{\partial\xi}-\theta'\frac{\partial f}{\partial\xi}\right] \tag{41}$$

The corresponding boundary conditions becomes.

$$\begin{aligned}
&f(0,\xi)=0,\ f'(0,\xi)=0,\ \phi'(0,\xi)=1,\ \theta(0,\xi)=1\\
&f'(\infty,\xi)=1,\ \phi'(\infty,\xi)=0,\ \theta(\infty,\xi)=0
\end{aligned} \tag{42}$$

It can be seen from equations (39)-(41) that for $\xi = 0.0$, the set of equations become similar by nature, solutions of which can easily be obtained by standard shooting method, otherwise these equations are locally non-similar, solution methodology of which will be discussed in the following sections. Once we know the solution of these equations, physical quantities of interest such as, the skin-friction coefficient, Cfx, and the magnetic intensity Mg_x, and the rate of heat transfer Nu_x at the surface may be calculated from

$$\begin{aligned}
Re_x^{1/2}Cf_x &= f''(\xi,0)\\
Re_x^{1/2}Mg_x &= -\phi(\xi,0)\\
Re_x^{-1/2}Nu_x &= -\theta'(\xi,0)
\end{aligned} \tag{43}$$

5.4. For small ξ

Since near the leading edge ξ is small ($\xi << 1$), solutions to the equations (39)-(41) with boundary conditions (42) may be obtained by using the perturbation method. We can expand all the depending functions in powers of ξ, we consider that

$$f(\xi,\eta) = \sum_{i}^{\infty} \xi^i f_i(\eta),\ \phi(\xi,\eta) = \sum_{i=0}^{\infty} \xi^i \phi_i(\eta),\ \theta(\xi,\eta) = \sum_{i=0}^{\infty} \xi^i \theta_i(\eta) \tag{44}$$

Substituting (44) into expression (39)-(41), and taking the terms only up to $O(\xi^2)$ we will get the system of equations together with boundary conditions (42) which is given as under:

$$f_0''' + \frac{1}{2}(f_0 U_0'' - S\phi_0\phi_0'') + \lambda\theta_0 = 0 \tag{45}$$

$$\frac{1}{Pm}\phi_0'' + \frac{1}{2}(f\phi_0' - \phi\phi_0') = 0 \tag{46}$$

$$\frac{1}{Pr}\theta_0'' + \frac{1}{2}f_0\theta_0' + f_0'\theta_0 = 0 \tag{47}$$

$$\begin{aligned} &f_0(0,\xi) = 0,\ f_0'(0,\xi) = 0,\ \phi_0'(0,\xi) = 1,\ \theta_0(0,\xi) = 1 \\ &f_0'(\infty,\xi) = 1,\ \phi_0'(\infty,\xi) = 0,\ \theta_0(\infty,\xi) = 0 \end{aligned} \tag{48}$$

$$f_1''' + \frac{1}{2}\left(f_0 f_1'' - f_0' f_1' - S(\phi_0\phi_1'' - \phi_0'\phi_1')\right) + (f_0'' f_1 - S\phi_0''\phi_1) + f_0'' + \lambda\theta_1 = 0 \tag{49}$$

$$\frac{1}{Pm}\phi_1'' + \frac{1}{2}(f_0\phi_1' - f_1'\phi_0) + (f_0'\phi_1 - f_1\phi_0') + \phi_0' = 0 \tag{50}$$

$$\frac{1}{Pr}\theta_1'' + \frac{1}{2}(f_0'\theta_1 + f_0\theta_1') + (f_1'\theta_0 + \theta_0' f_1) + \theta_0' = 0 \tag{51}$$

$$\begin{aligned} &f_1(0,\xi) = 0,\ f_1'(0,\xi) = 0,\ \phi_1'(0,\xi) = 0,\ \theta_1(0,\xi) = 0 \\ &f_1'(\infty,\xi) = 0,\ \phi_1'(\infty,\xi) = 0,\ \theta_1(\infty,\xi) = 0 \end{aligned} \tag{52}$$

$$\begin{aligned} &f_2''' + \frac{1}{2}\left(f_0 f_2'' - f_1'^2 - S(\phi_0\phi_2'' - \phi_1'^2)\right) + \left(f_1 f_1'' - f_0' f_2' - S(\phi_1\phi_1'' - \phi_0'\phi_2')\right) \\ &+ \frac{3}{2}(f_0'' f_2 - S\phi_0''\phi_2) + f_1'' + \lambda\theta_2 = 0 \end{aligned} \tag{53}$$

$$\frac{1}{Pm}\phi_2'' + \frac{1}{2}(f_0\phi_2' - f_0'\phi_2) + (f_1\phi_1' - f_1'\phi_1) + \frac{3}{2}(\phi_0' f_2 - f_0'\phi_2) + \phi_1' = 0 \tag{54}$$

$$\frac{1}{Pr}\theta_2'' + \frac{1}{2}(f_0\theta_2' + \theta_1 f_1') + (\theta_1' f_1 + f_2'\theta_0) + \frac{3}{2}f_2\theta_0' + \theta_1' = 0 \tag{55}$$

$$\begin{aligned} &f_2(0,\xi) = 0,\ f_2'(0,\xi) = 0,\ \phi_2'(0,\xi) = 0,\ \theta_2(0,\xi) = 0 \\ &f_2'(\infty,\xi) = 0,\ \phi_2'(\infty,\xi) = 0,\ \theta_2(\infty,\xi) = 0 \end{aligned} \tag{56}$$

It is pertinent to mentioned that the equations (45) -(56) are coupled and non-linear, so the solutions of these equations are obtained by the Nachtsheim-Swigert iteration technique together with the sixth order implicit Runge-Kutta-Butcher initial value solver. After knowing the values of the functions $f''(\eta)$, $\phi(\eta)$ and $\theta'(\eta)$ and their derivatives we can calculate the values of the coefficient of skin friction, surface magnetic intensity and heat transfer in the region near the leading edge against ξ from the following expansion for S=0.8, λ=0.5, Pm =0.1, Pr=0.1 respectively.

$$\begin{aligned} Re_x^{1/2}Cf_x &= (1.06920 + 1.419350\xi + 0.410048\xi^2 + ...) \\ Re_x^{1/2}Mg_x &= -(3.535877 + 1.024740\xi - 0.508466\xi^2 + ...) \\ Re_x^{-1/2}Nu_x &= -(-0.06887 - 0.149626\xi + 0.043242\xi^2.....) \end{aligned} \tag{57}$$

The numerical results $Re_x^{1/2}Cf_x$, $Re_x^{1/2}Mg_x$ and $Re_x^{-1/2}Nu_x$ thus obtained are entered in tables 1, 2 and 3. We can see that from these tables the asymptotic solutions are smaller values of the local transpiration parameter $\xi\ \varepsilon[0, 1]$ are found in excellent agreement with that obtained from free variable formulation by finite difference.

In Table 4 the comparison of the solutions obtained by finite difference method and that Glauert[11] obtained analytically has been shown for the coefficient of skin friction, $f''(0)$ and the coefficient of magnetic intensity at the surface, $\phi(0)$. It is also observed that for $Pm = 1.0, 10.0$ and the variation in magnetic field parameter S decrease the skin friction coefficient, $f''(0)$ which also approaches to zero as S approaches to 1.1. It can further be seen that value of $\phi(0)$ increases due to increase of magnetic field parameter S. From this table we also found that the present numerical solutions and the analytical solutions obtained by Glauert[11] are in good agreement that is for magnetic Prandtl number Pm=10.0. Here, we also see that for smaller value of magnetic Prandtl number, (say) Pm=0.1, and for magnetic field parameter $S = \pi$ flow separation occurs which had also been observed by Glauert[11].

Present numerical values of $f''(0)$ for magnetic Prandtl number Pm in [0.1, 100] while S=0.1 and 0.05 are entered in Table 5 and compared with that obtained by Glauert[11] and Davies[2], respectively, for smaller and larger value of Pm and found to be in good agreement. Finally, the results entered in Table 6 are those obtained for heat transfer coefficient, $\theta'(0)$ by the present method and of Ramamoorthy[6] for comparison. From this table it can be seen that the numerical values of $\theta'(0)$ we have obtained for different values of S = 0.1, 0.3 and 0.5 while Pm=0.1, Pr=1.0, λ=0.5 are very closed to that of Ramamoorthy [6].

5.5. For large ξ

Now, attention is given in finding the solution of equations (39)-(42) when ξ is large. The order of magnitude analysis of various terms in these equations shows that U''' and $\xi U''$ are largest terms in equation (39) and ϕ'' and $\xi\phi'$ in equation (40) and θ'' and $\xi\theta'$ in equation (41). In the respective equations, both the terms have to be balanced in magnitude and the only way to do this, is to assume that η is small and hence its derivative is large. It is essential to find appropriate scaling for U, ϕ and θ. On balancing U''' and $\xi U''$ in equation (39) and ϕ'' and $\xi\phi'$ in equation (40) and θ'', $\xi\theta'$ in equation (41), it is found that $\eta = O(\xi^{-1})$, $U = O(\xi^{-1})$, and $\phi = O(\xi^{-1})$. Therefore, following transformations may be introduced.

$$U(\eta) = \xi^{-1}\bar{f}(\bar{\eta}),\ \bar{\eta} = \xi\eta,\ \phi(\eta) = \xi^{-1}\bar{\phi}(\bar{\eta}),\ \theta(\eta) = \bar{\theta}(\bar{\eta}) \tag{58}$$

By using (57), the transformed equation will take the form:

$$\bar{f}''' + \bar{f}'' + \lambda\xi^{-2}\theta = \frac{1}{2}\left[\bar{f}'\frac{\partial \bar{f}}{\partial \xi} - \bar{f}''\frac{\partial \bar{f}}{\partial \xi} - S\left(\bar{\phi}'\frac{\partial \bar{\phi}'}{\partial \xi} - \bar{\phi}''\frac{\partial \bar{\phi}}{\partial \xi}\right)\right] \quad (59)$$

$$\frac{1}{Pm}\bar{\phi}'' + \bar{\phi}' = \frac{1}{2}\xi^{-1}\left[\bar{U}'\frac{\partial \bar{\phi}}{\partial \xi} - \bar{f}'\frac{\partial \bar{f}}{\partial \xi}\right] \quad (60)$$

$$\frac{1}{Pr}\bar{\theta}'' + \bar{\theta}' = \frac{1}{2}\xi^{-1}\left[\bar{f}'\frac{\partial \bar{\theta}}{\partial \xi} - \bar{\theta}'\frac{\partial \bar{U}}{\partial \xi}\right] \quad (61)$$

The regular perturbation of the functions $\bar{f}$, $\bar{\phi}$ and $\bar{\theta}$ in power of ξ^{-2} is given as under:

$$\bar{f}(\xi,\bar{\eta}) = \sum_{m=0}^{1}\xi^{-2m}\bar{f}_m(\bar{\eta}),\ \bar{\phi}(\xi,\bar{\eta}) = \sum_{m=0}^{1}\xi^{-2m}\bar{\phi}_m(\bar{\eta}),\ \bar{\theta}(\xi,\bar{\eta}) = \sum_{m=0}^{1}\xi^{-2m}\bar{\theta}_m(\bar{\eta}) \quad (62)$$

By substituting (61) into (58)-(60), and equating like power of ξ and by dropping bars we have the following set of equations:
$O(\xi^0)$:

$$f_0''' + f_0'' = 0 \quad (63)$$

$$\frac{1}{Pm}\phi_0'' + \phi_0' = 0 \quad (64)$$

$$\frac{1}{Pr}\theta_0'' + \theta_0' = 0 \quad (65)$$

The boundary conditions regarding to the perturbation series expansion are of the following form:

$$\begin{gathered} f_0(0) = 0,\ f_0'(0) = 0,\ \phi_0'(0) = 1,\ \theta_0(0) = 1 \\ f_0'(\infty) = 1,\ \phi_0'(\infty) = 0,\ \theta_0(\infty) = 0 \end{gathered} \quad (66)$$

$O(\xi^{-2})$:

$$f_1''' + f_1'' + \lambda\theta_0 = 0 \quad (67)$$

$$\frac{1}{Pm}\phi_1'' + \phi_1' = 0 \quad (68)$$

$$\frac{1}{Pr}\theta_1'' + f_0'\theta_0 + \theta_1' = 0 \quad (69)$$

And the boundary conditions are as follows:

$$\begin{gathered} f_1(0) = 0,\ f_1'(0) = 0,\ \phi_1'(0) = 0,\ \theta_1(0) = 0 \\ f_1'(\infty) = 0,\ \phi_1'(\infty) = 0,\ \theta_1(\infty) = 0 \end{gathered} \quad (70)$$

From the solutions of (62)-(65) and (66)-(69) we obtain

$$f_0(\eta) = \eta + e^{-\eta} - 1 \quad (71)$$

$$\phi_0(\eta) = -\frac{1}{Pm}e^{-\eta Pm} \quad (72)$$

$$\theta_0'(\eta) = Pre^{-\eta Pr} \tag{73}$$

$$f_1''(\eta) = \frac{\lambda e^{-\eta Pr}(Pr-1) + Pr\lambda e^{-\eta} - \lambda e^{-\eta}}{2(Pr-1)} \tag{74}$$

$$\phi_1(\eta) = 0 \tag{75}$$

$$\theta_1'(\eta) = 0 \tag{76}$$

Now, knowing the values of $f_0''(0)$, $\phi_0(0)$, $\theta_0'(0)$, $f_1''(0)$ $\phi_1(0)$and $\theta_1'(0)$ from the above solutions we calculate the friction coefficient, $R_{ex}^{\frac{1}{2}}Cf_x$, local rate of heat transfer, $R_{ex}^{-\frac{1}{2}}Nu_x$, and the local magnetic intensity, $R_{ex}^{\frac{1}{2}}Mg_x$ at the surface from the following expressions:

$$\begin{aligned} Re_x^{1/2}Cf_x &\simeq \xi + \frac{\lambda(Pr-1) + Pr\lambda - \lambda}{2(Pr-1)}\xi^{-1} \\ Re_x^{1/2}Mg_x &\simeq \frac{\xi}{Pm} \\ Re_x^{-1/2}Nu_x &\simeq Pr\xi \end{aligned} \tag{77}$$

The asymptotic values at large values of transpiration parameter ξ of t$Re_x^{1/2}Cf_x$, $Re_x^{1/2}Mg_x$ and $Re_x^{-1/2}Nu_x$ obtained from the relations (77) for different values of magnetic field parameter S, magnetic Prandtl number Pm and the Prandtl number Pr respectively are entered, respectively, in tables 1, 2 and 3. From these tables it can be seen that at larger values of ξ the value of $Re_x^{1/2}Cf_x$ approaches to ξ, that of $Re_x^{1/2}Mg_x$ and approaches, respectively, to ξ/Pm and $Pr\xi$. Further to accept that these values are in excellent agreement with that obtained numerically by the finite difference method.

6. Conclusion

The physical parameters such as mixed convection parameter λ, transpiration parameter ξ, magnetic field parameter S, magnetic Prandtl number Pm and Prandtl number Pr exerts significant influence on coefficients of skin friction $Re_x^{1/2}Cf_x$, heat transfer $Re_x^{-1/2}Nu_x$ and magnetic intensity $Re_x^{1/2}Mg_x$ at the surface.

- The coefficients of skin friction, heat transfer increases and magnetic intensity at the surface decreases with the increase of mixed convection parameter λ. The momentum and thermal boundary layer thicknesses decreases and velocity and temperature profiles increases with the increase of the mixed convection parameter λ. It is also noted that the increase in mixed convection parameter λ reduce the transverse component of magnetic field profile.
- It is very interesting fact that the coefficient of skin friction, rate of heat transfer decreases and magnetic intensity at the surface increases with the increase of magnetic field parameter S. The momentum and thermal boundary layer thicknesses increases with the increase of magnetic field parameter S and the transverse component of magnetic field also increases with the increase of magnetic field parameter S.

• The transpiration parameter ξ play a significant role in boundary layer, due to increase in transpiration parameter ξ the momentum and thermal boundary layer thicknesses decreases and the transverse component of magnetic field profile also reduced.

• The coefficient of skin friction, heat transfer increases and the coefficient of magnetic intensity decreases with the increase of magnetic Prandtl number Pm.

• The increase in Prandtl number Pr reduce the coefficient of skin friction and enhance the coefficient of heat transfer and magnetic intensity at the surface.

References

[1] H. P. Greenspan and G. F. Carrier, The magnetohydrodynamic flow past a flat plate, J. Fluid Mech. 6, 77-96 (1959).

[2] T. V. Davies, The magnetohydrodynamic boundary layer in two-dimensional steady flow past a semi-infinite flat plate, Part I, Uniform conditions at infinity, Proc. R. Soc. Lond. A 273, 496-507 (1963).

[3] T. V. Davies, The magnetohydrodynamic boundary layer in two-dimensional steady flow past a semi-infinite flat plate, Part III, Influence of adverse magneto-dynamic pressure gradient, Proc. R. Soc. Lond. A 273, 518-537 (1963).

[4] R. J. Gibben, Magnetohydrodynamic stagnation-point flow, Quart. J. Mech. and Appled Math, VoL XVIII, Pt. 3, (1963)

[5] R. J. Gibben, Magnetohydrodynamic boundary layers in presence of pressure gradient, Proc. R. Soc. Lond. A 287, 123-141 (1965)

[6] P. Ramamoorthy, Heat transfer in hydromagnetics, Q. J. Mech. Appl. Math 18, 31-40 (1965).

[7] C. W. Tan and C. T. Wang, Heat transfer in aligned-field magnetohydrodynamic flow past a flat plate, Int. J. Heat Mass Transf. 11, 319-329 (1967).

[8] T. Hildyard, Falkner-Skan Problem in Magnetohydrodynamics, Phys. Fluids 15, 1023-1027 (1972).

[9] S. S. Chawla, Magnetohydrodynamic oscillatory flow past a semi-infinite flat plate, Int. J. Non-Linear Mech. 6, 117-134 (1971).

[10] D. B. Ingham, The magnetogasdynamic boundary layer for a thermally conducting plate, Quart. J. Mech. and Applied Math., Vol. XX, Pt. 3, 1967

[11] Glauert, M. B. J. Fluid Mech. 12(1962), 625.

[12] S. S. Chawla,Fluctuating boundary layer on a magnetized plate,Proc. Comb. Phil.Soc. (1967), 63, 513.

[13] H. T. Lin and L. K. Lin, Similarity solutions for laminar forced convection heat transfer from wedges to fluids of any Prandtl number, Int. J. Heat Mass Transf. 30, 1111-1118 (1987).

[14] J. C. Y. Koh and J. P. Hartnett, Skin-friction and heat transfer for incompressible laminar flow over porous wedges with suction and variable wall temperature, Int. J. Heat Mass Transf. 2, 185198 (1961).

[15] M. A. Hossain. K. Mahmood, Effect of transpiration on hydromagnetic flow and heat transfer, Journal of Bangladesh academy of Sciences, Vol. 17, No.1 11-20-1993.

[16] N.G. Kafoussias and N. D. Nanousis, Magnetohydrodynamic laminar boundary-layer flow over a wedge with suction or injection, Can. J. Phys. 75, 733-745 (1997).

[17] M. Mahmood, S. Asghar, and M. A. Hossain, Hydromagnetic flow of viscous incompressible fluid past a wedge with permeable surface, ZAMM. Z. Angew. Math. Mech. 89, No.3. 174-188 (2009).

INDEX

D

E

F

G

H

I

J

K

L

M

N

O

P

U

V

W

Y

Z